# Energy Management of Prosumer Communities

# Energy Management of Prosumer Communities

Editors

**Seppo Sierla**
**Mahdi Pourakbari-Kasmaei**

MDPI • Basel • Beijing • Wuhan • Barcelona • Belgrade • Manchester • Tokyo • Cluj • Tianjin

*Editors*
Seppo Sierla
Electrical Engineering and Automation
Aalto University
Espoo
Finland

Mahdi Pourakbari-Kasmaei
Electrical Engineering and Automation
Aalto University
Espoo
Finland

*Editorial Office*
MDPI
St. Alban-Anlage 66
4052 Basel, Switzerland

This is a reprint of articles from the Special Issue published online in the open access journal *Energies* (ISSN 1996-1073) (available at: www.mdpi.com/journal/energies/special_issues/Prosumer_Communities).

For citation purposes, cite each article independently as indicated on the article page online and as indicated below:

| LastName, A.A.; LastName, B.B.; LastName, C.C. Article Title. *Journal Name* **Year**, *Volume Number*, Page Range. |
|---|

**ISBN 978-3-0365-2058-2 (Hbk)**
**ISBN 978-3-0365-2057-5 (PDF)**

# Contents

# About the Editors

**Seppo Sierla**

Seppo Sierla received his Ph.D. degree in information technology in automation from Helsinki University of Technology in 2007 and his title of docent in the field of "Software Design for Industrial Automation" from Aalto University in 2013. As a docent, he has been a principal investigator or key expert in numerous project related to the application of machine learning, simulation and digital twin technologies in the energy, manufacturing and process industry sectors. Currently, he is a university lecturer with the Department of Electrical Engineering and Automation, Aalto University, Finland. He is the treasurer of the IEEE Finland Section and the conference treasurer, keynote co-chair or exhibitions co-chair of IEEE conferences organized at the department: INDIN2019, MLSP2020, ISGT2021 Europe, ISIT2022 and ISIE2023. He has served as a guest editor for several special issues in the journals *Energies, Applied Sciences* and *Frontiers in Energy Research.*

**Mahdi Pourakbari-Kasmaei**

Mahdi Pourakbari-Kasmaei received his Ph.D. degree in electrical engineering, power systems from the São Paulo State University (UNESP), Ilha Solteira, Brazil, in 2015. He was a postdoctoral fellow at UNESP and a visiting researcher at the University of Castilla-La Mancha, Spain. He was a project executive or principal investigator of several practical and academic projects and a consultant in an electric power distribution company. Currently, he is an assistant pProfessor with the Department of Electrical Engineering and Automation, Aalto University, Finland. He is the chairman of the IEEE PES Finland IE13/PE31/34/PEL35 Joint Chapter, the general chair of the IEEE PES ISGT-Europe 2021 Conference and an associate editor of several journals such as *IEEE Access; Journal of Control, Automation and Electrical Systems*; etc. His research interests include power systems planning, operations, economics, environmental issues, and power system protection and transients.

*energies*

MDPI

*Article*

# A Simulation Environment for Training a Reinforcement Learning Agent Trading a Battery Storage

**Harri Aaltonen [1,*], Seppo Sierla [1], Rakshith Subramanya [1] and Valeriy Vyatkin [1,2,3]**

[1] Department of Electrical Engineering and Automation, School of Electrical Engineering, Aalto University, FI-00076 Espoo, Finland; seppo.sierla@aalto.fi (S.S.); rakshith.subramanya@aalto.fi (R.S.); valeriy.vyatkin@aalto.fi (V.V.)

[2] Department of Computer Science, Electrical and Space Engineering, Luleå University of Technology, 97187 Luleå, Sweden

[3] International Research Laboratory of Computer Technologies, ITMO University, 197101 St. Petersburg, Russia

* Correspondence: harri.aaltonen@aalto.fi

**Abstract:** Battery storages are an essential element of the emerging smart grid. Compared to other distributed intelligent energy resources, batteries have the advantage of being able to rapidly react to events such as renewable generation fluctuations or grid disturbances. There is a lack of research on ways to profitably exploit this ability. Any solution needs to consider rapid electrical phenomena as well as the much slower dynamics of relevant electricity markets. Reinforcement learning is a branch of artificial intelligence that has shown promise in optimizing complex problems involving uncertainty. This article applies reinforcement learning to the problem of trading batteries. The problem involves two timescales, both of which are important for profitability. Firstly, trading the battery capacity must occur on the timescale of the chosen electricity markets. Secondly, the real-time operation of the battery must ensure that no financial penalties are incurred from failing to meet the technical specification. The trading-related decisions must be done under uncertainties, such as unknown future market prices and unpredictable power grid disturbances. In this article, a simulation model of a battery system is proposed as the environment to train a reinforcement learning agent to make such decisions. The system is demonstrated with an application of the battery to Finnish primary frequency reserve markets.

**Keywords:** battery; reinforcement learning; simulation; frequency reserve; frequency containment reserve; timescale; artificial intelligence; real-time; electricity market



**Citation:** Aaltonen, H.; Sierla, S.; Subramanya, R.; Vyatkin, V. A Simulation Environment for Training a Reinforcement Learning Agent Trading a Battery Storage. *Energies* **2021**, *14*, 5587. https://doi.org/10.3390/en14175587

Academic Editor: Branislav Hredzak

Received: 8 July 2021
Accepted: 12 August 2021
Published: 6 September 2021



## 1. Introduction

Battery storages are an essential element of the emerging smart grid. Batteries are crucial for coping with increased photovoltaic [1] and wind penetration [2]. Schemes for introducing batteries are proposed at the level of buildings [3], wind farms [4] and the distribution grid [5]. Electric vehicle batteries can be used to temporarily store excess rooftop photovoltaic generation, which can be used to supply the load after photovoltaic generation has dropped [6]. Significant recent research has emerged on reinforcement learning (RL) applications for complex decision-making involving battery systems and energy markets. However, such works frequently ignore short-term electrical phenomena and employ RL frameworks with the simplifying assumption that renewable generation, power consumption and battery charging and discharging power remain constant throughout each market interval. Such assumptions are usually captured by a set of equations that specifies the environment of the RL agent. The environment is a system for interactive training of an RL agent: when the agent takes actions such as placing bids on a market, the environment gives feedback about the beneficial as well as the undesirable outcomes resulting from the action. If these simplifying assumptions could be eliminated, RL-powered battery systems could be a solution for managing short-term phenomena such as fluctuating renewable generation and power consumption, as well as sudden grid disturbances. To this end, this article presents an RL application working on two timescales: the timescale of the markets and the short-term timescale of electrical phenomena.

There are numerous applications for quickly reacting batteries. For example, batteries can support the extraction of maximum power generation from photovoltaic batteries with real-time maximum power point tracking control [7,8]. Battery applications for smoothing fluctuations of wind power generation require real-time control [9,10]. Without such smoothing applications, grid operations are required to take countermeasures to manage the resulting grid frequency variations [11]. One way to use batteries is to directly react to such frequency variations. Frequency reserves are energy resources that stand by to react to such frequency deviations by adjusting their production and consumption. Depending on the region, transmission system operators (TSO) or independent system operators (ISO) operate frequency reserve markets in which they procure frequency reserves and pay compensations for the provider of the reserve resource. Out of the various frequency reserve markets, primary frequency reserves (PFR) have the fastest response time requirements, which is reflected in the financial compensations paid to the reserve resource providers [12]. As batteries are easily capable of meeting such requirements, PFR participation allows batteries to contribute to coping with imbalances in the grid, regardless of whether such imbalances are caused by fluctuations in photovoltaic or wind generation, changes in electricity consumption or other disturbances [13].

PFR markets are generally auctions, in which the provider of the reserve resource has to specify the reserve capacity (adjustable MW of power production or consumption) for each market interval of the upcoming bidding period. A common market structure is that the bidding period is day-ahead and that the interval is one hour; this is also the case in the Finnish PFR market Frequency Containment Reserve for Normal Operation (FCR-N) [14], which will be the case study of this paper. Although revenues can be increased by bidding on as many intervals as possible, and with as much capacity as possible, the market will penalize participants that fail to provide the capacity specified in their bid. In the case of a battery storage, such failures will occur whenever the battery state of charge (SoC) reaches a minimum or maximum limit. As PFR requires the battery to react to frequency deviations on the order of seconds, it is an application operating on the two timescales identified above: the timescale of the markets and the short-term timescale of electrical phenomena. The contribution of this paper is an RL solution operating on these two timescales and using a simulation model to accurately capture the dynamics of the battery. The RL agent bids on the PFR market, and its training environment is a simulation model in which the battery reacts to grid frequency deviations with a one-second time step.

This paper is structured as follows: Section 2 reviews the state of the art. Section 3 presents a semiformal description of the solution. Section 4 describes the implementation of the simulation as well as the RL, with an application to the Northern European PFR market. Section 5 presents results of running the RL bidder on this market. Section 6 concludes the paper with an assessment of the obtained results and a discussion of further research directions.

## 2. Literature Review

There is a lack of research on using RL to trade batteries on PFR markets. However, there is a growing body of research on RL applications for batteries. There is also research on battery applications for frequency regulation.

### 2.1. Batteries in Primary Frequency Reserves

The increased reliance on renewable generation [15] and unreliabilities resulting from a rapid drive towards a smart grid [16] are increasing the demand for PFR, which has traditionally been provided by fossil fuel-based solutions [17]. There is a growing volume of research on solutions for providing PFR with distributed intelligent energy resources such as electric vehicles [18,19], domestic loads [20,21] and industrial processes [22]. The increased penetration of renewables is driving investments to battery-provided PFR [23]. However, since batteries can be exploited for a variety of grid support services, their penetration in PFR markets will depend on how prices on these markets develop [24]. Although batteries have been economically viable PFR assets for a long time [25], a growing body of research has emerged only in the last few years. The economic disadvantages caused by the battery degradation resulting from PFR participation are well understood and do not prevent economically profitable PFR participation [26,27]. Srinivasan et al. [28] propose the use of a virtual power plant to complement batteries with other intelligent distributed energy resources providing PFR.

### 2.2. Reinforcement Learning Applications for Batteries

In this section, RL applications for batteries are reviewed according to the timescale in which they operate. Three distinct timescales have been identified:

- Real-time control.
- Medium-term decision-making for optimizing some operational criteria such as electricity costs or photovoltaic self-consumption. In many cases, this involves decision-making once per electricity market interval, which in many cases is hourly.
- Long-term studies to support investment decisions.

RL is broadly applied in real-time control, and research exists for a variety of battery applications. Maximizing photovoltaic generation requires a control algorithm such as mean power point tracking [29]. Real-time control with respect to driving speed is required for advanced battery management applications in electric vehicles [30] and plug-in hybrid electric vehicles (Chen et al. [31]). As a semi-real-time example, Sui and Song [32] used RL with a one-minute timestep to manage battery temperatures and thus battery lifetime in a battery pack.

Medium-term RL applications adjust the parameters of intelligent battery systems to optimize their operation. Muriithi and Chowdhury [33] optimized a battery and local photovoltaic to minimize electricity bills under variable electricity prices. Batteries have been used in conjunction with reschedulable loads to perform the rescheduling to exploit time-of-use and real-time energy pricing schemes [34–36] and variable intraday electricity market prices [37]. Whereas most works are aimed at existing electricity markets, a few authors have demonstrated the benefits of RL to optimize the emerging decentralized electricity system on novel markets [38,39]. The above works involved decision-making on electricity markets, which is the most common type of RL application in this category. However, other kinds of applications also exist. Mbuwir et al. [40] used a battery to maximize self-consumption of a local photovoltaic system. Finally, it is noted that for building HVAC systems, a thermal energy storage can be a competitor to a battery storage [41].

RL can be used at investment time to determine the parameters of a smart energy system that incorporates batteries. Diverse application contexts have been encountered, including wireless EV charging systems [42], wind farms [43], microgrids [44] and isolated villages with microgrids [45].

There is a lack of RL applications combining multiple timescales. In particular, the referenced works addressing the timescale of relevant markets ignore phenomena requiring real-time control actions, so there is a lack of research on how to financially exploit RL applications for batteries that try to solve the global problem of smoothing the fluctuations of renewable power generation. In this article, an RL agent is presented for trading on hourly PFR markets, so that the impact of power grid frequency fluctuations is considered on the timescale of seconds.

## 3. Battery Trading System

Figure 1 presents an overview of the proposed system. The bidding agent is implemented with a neural network, and it operates on the timescale of the market. The environment includes an offline implementation of the frequency market, based on market data, as well as a real-time battery simulation, which will detect if the battery goes out of bounds. The simulation has two modes: a PFR market participation mode and a resting mode, for hours with no PFR participation, during which the battery state of charge is driven to a value that is ideal with respect to upcoming PFR participation. Two items of state information are provided by the environment to the RL agent: battery state information and a market forecast. While battery state of charge information could be very useful for the RL, it would not be available at the timeframe when the bidding is done on the previous day. Thus, the only battery state information that is available is the information of when the battery last rested—this information can readily be extracted from the bidding plans. The PFR market forecast is also relevant, since it may be beneficial to concentrate bidding on high-price hours and resting on low-price hours. The day-ahead PFR forecasting method presented in [14] is used.

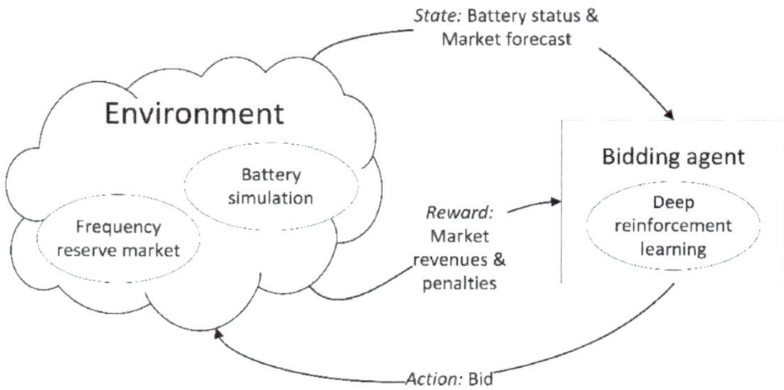


**Figure 1.** System overview.

Several formulations of the reinforcement learning mechanism exist, and these have been applied in the battery energy management domain. The simplest is q-learning, which involves a table for mapping states and actions [46]. As our problem formulation involves a small state space, q-learning could have been used in this work. However, the use of q-learning would have introduced scalability problems in further work involving more complex state spaces. Reinforcement learning methods using a neural network instead of a q-table are a more scalable approach. Such methods are called deep reinforcement learning in case there is more than one hidden layer [30]. In our case, a neural network with one hidden layer was used, since experimentation with a second hidden layer did not result in improved performance. Advanced variations of deep reinforcement learning involve the use of several interdependent neural networks. This is a beneficial approach when the state space becomes significantly more complicated, as in the case of Zhang et al. [47] who consider a system with several resources in each of the following categories: batteries wind and photovoltaic generation, water purification plants and diesel generators.

In order to support a problem formulation of the concept in Figure 1, Table 1 defines relevant symbols and Table 2 defines functions, which are used by the algorithm for training the reinforcement learning agent. Figure 2 formalizes the concept in Figure 1 using these symbols. Figure 3 presents the algorithm for training the reinforcement learning agent ('bidding agent') in the environment of Figure 2. The algorithm is based on established reinforcement learning techniques and integrates a real-time simulation of the battery on a primary frequency reserves market. A time range of days is selected for the training. One epoch is one iteration of the outer loop in Figure 3 and involves running the agent for each day in the training period. One state–action pair of the reinforcement learning agent is one hour, since that is the primary frequency reserves market interval. One state–action pair is taken by one iteration of the inner loop of the unshaded area in Figure 3 (i.e., the loop with the condition '$h < 24$', which iterates through each hour of the day). The shaded area of Figure 3 involves calling the battery simulation with a one-second timestep. The purpose of this is to determine whether the battery state of charge goes out of bounds which involves a penalty from the primary frequency reserves market, since the battery is not available to provide the primary frequency reserves in such a state.

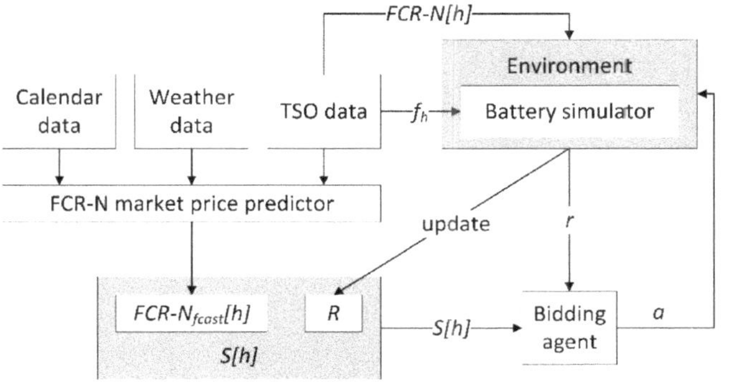


**Figure 2.** Setup for training the bidding agent.

**Table 1.** Symbols.

| Symbol | Data Type | Description |
|---|---|---|
| *SoC* | Float | State of charge of the battery expressed as percentage of full charge |
| $OoB_{min}$ | Boolean | True if (*SoC*) was out of bounds (*OoB*) at any time during a specific minute |
| $OoB_s$ | Boolean | True if *SoC* was out of bounds for a one-second timestep of the battery simulation |
| *R* | Integer | Hours since the battery last rested. Resting is defined as not participating in the frequency reserve market and charging/discharging to bring the *SoC* to 50%. E.g., if the battery rested most recently on the previous hour, *R = 0* |
| *day* | Date | The current day corresponding to the current state of the environment |
| $day_{start}$ | Date | The first day of the training set |
| $day_{end}$ | Date | The last day of the training set |
| *h* | Integer in range 0–23 | The current hour (the current day is stored in the symbol *day*) |
| $FCR\text{-}N_{fcast}[h]$ | Float | The forecasted FCR-N market price in EUR per megawatt (EUR/MW) for hour *h* |
| *FCR-N[h]* | Float | The actual FCR-N market price (EUR/MW) for hour *h* |
| *S[h]* | [Integer, Float] | State of the environment at the hour *h*, i.e., $[R, FCR\text{-}N_{fcast}[h]]$ |
| $f_h$ | Float [3600] | Power grid frequency time series for hour *h*. One data point per second |
| *a* | Integer in range 0–3 | Action to be taken by the bidding agent. 3 = rest (no bid), 2 = bid with 600 kilowatt (kW) capacity, 1 = bid with 800 kW capacity, 0 = bid with 1 MW capacity |
| *r* | float | Reward |
| *trace* | Array with elements of type *[S[h], a, r, S[h + 1]]* | An experience trace consisting of all of the experiences collected during one epoch. A single experience consists of the following: *[S[h], a, r, S[h + 1]]* |
| *maxEp* | Integer | The maximum number of epochs used to train the reinforcement learning agent |
| $penalty_{min}$ | Integer | The number of minutes during the current hour in which the battery was not available for providing frequency reserves and thus incurred a financial penalty from the frequency reserve market |
| *compensation[h]* | float | The compensation in EUR for participating in FCR-N for the hour h |
| *penalty[h]* | float | The penalty in EUR for the battery being unavailable while participating in FCR-N for the hour h |
| $reputation_{damage}$ | float | A quantification in EUR of the damage to the reputation of the FCR-N reserve provider (i.e., the battery operator), due to failures to provide the reserve |
| $reputation_{factor}$ | float | A coefficient in EUR that can be adjusted to train the bidding agent to avoid penalties |

**Table 2.** Functions used in the procedure for training the reinforcement learning agent (Figure 3).

| Function | Description |
|---|---|
| *reset(S*[0]) | Resets the state variables at the beginning of the epoch |
| *pow = ctrl(a)* | The parameter *a* is the frequency reserve capacity in kW that the bidding agent decided to bid on the frequency reserve market. The output *pow* is the discharge/charge power command to the battery from the battery controller. The output is determined according to the frequency data from $f_h$ and the FCR-N market technical specification [48]. |
| $OoB_s = sim(pow)$ | This function runs the battery simulation for one second, according to the *pow* charge/discharge command from *ctrl(a)*. *SoC* is an internal state variable of the battery simulator. If the *SoC* goes OoB during this second, the $OoB_s$ output is true, otherwise false. |
| $OoB_{min} = bounds(OoB_s)$ | If $OoB_s$ is true, $OoB_{min}$ is set true. Otherwise, no action. |
| *capacity(a)* | The capacity in MW of the bid corresponding to the action a taken by the agent. The capacity is 0 if a = 3, 0.6 MW if *a* = 2, 0.8 MW if *a* = 1 and 1.0 MW if *a* = 0. |
| *S[h] = state(h)* | Construct the state data structure *S[h]* with the current value of *R* and *h*. |

Finally, the environment needs to provide feedback to the RL agent in the form of a reward. The compensation from the PFR markets and the penalties for failing to provide the reserve are elements of the reward. For a particular hour, the compensation from the market is the product of the market price EUR/MW and the reserve capacity in MW. The FCR-N technical specification states that the compensation is only received for those minutes when the reserve was available [49]:

$$compensation = \frac{60 - penalty_{min}}{60} FCR\_N[h] \times capacity(a), \quad (1)$$

For each hour, the compensation is paid only for those minutes during which the system did not violate the penalty criteria. For this reason, Equations (1) and (2) include the fraction (60_penalty_min)/60 [49]:

$$penalty = \frac{penalty_{min}}{60} FCR_{N[h]} \times capacity(a), \quad (2)$$

The terms and conditions for providers of FCR state that if the reserve resource is unavailable too often, the frequency reserve market operator may, at its discretion, temporarily ban the reserve provider from participating in the market [50]. In order to include such considerations in the learning process of the reinforcement learning agent, a $reputation_{damage}$ is defined. This differs from the penalty in Equation (2) in two respects. Firstly, it is not dependent on the FCR-N price for the hour in question [50]. Secondly, since the market operator does not provide any quantitative criteria for banning the reserve provider [50], a $reputation_{factor}$ coefficient is defined, which can be adjusted by the reserve provider in order to make the tradeoff between increasing revenues versus bidding prudently to avoid penalties:

$$reputation_{damage} = reputation_{factor} \times \frac{penalty_{min}}{60} \times capacity(a), \quad (3)$$

Therefore, the reward for the reinforcement learning agent is the formula for the net revenue with an additional element to further penalize the agent for failing to provide the reserve and thus damaging the reputation of the reserve provider:

$$r = compensation - penalty - reputation_{damage}, \quad (4)$$

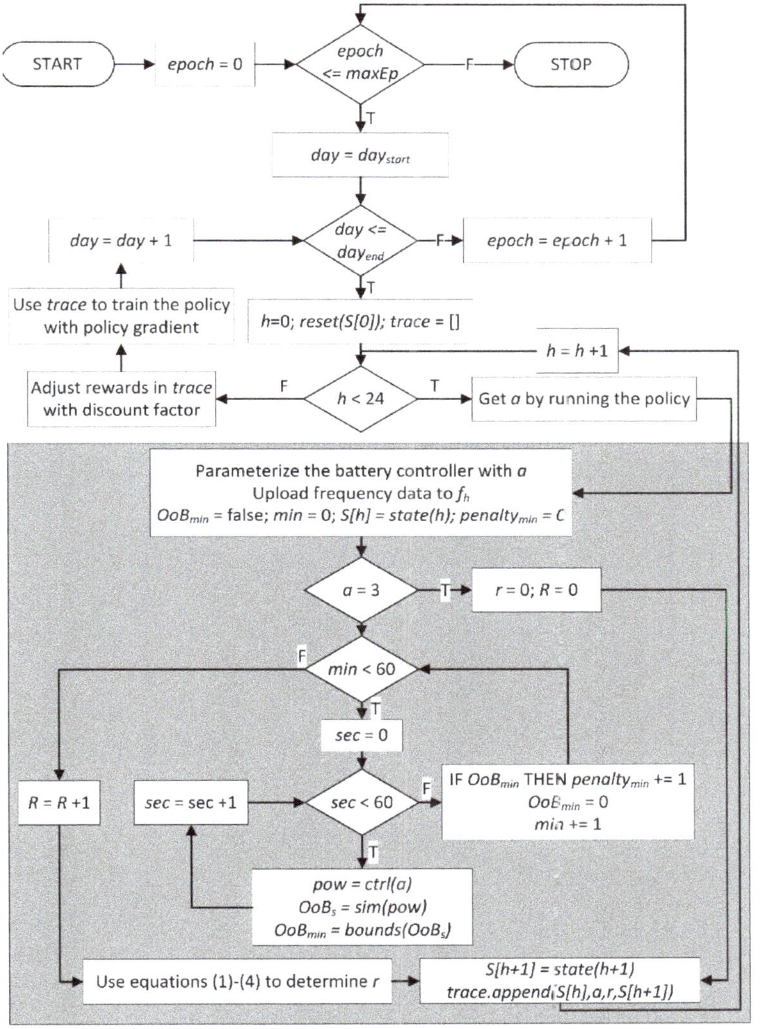


**Figure 3.** Procedure for training the bidding agent using the environment in Figure 2. The shaded area is the step (a) function discussed in Section 4.1.

## 4. Implementation

### *4.1. Enviroment*

The battery model in Figure 4 is used to simulate the behavior of the battery's SoC when it is charged or discharged as it participates on PFR. The battery model is a Simulink model and receives its inputs from the MATLAB function that implements the *ctrl(a)* function in Table 2. The implementation is done according to the rules of the Finnish PFR market FCR-N [48]. The same rules apply to PFR markets in Sweden, Norway and Denmark. In these countries, the nominal power grid frequency is 50 Hz with a maximum permitted deadband zone when the grid frequency is in the range 49.99–50.01 Hz. Equation (5) defines the discharging power when the frequency is in the range 49.9–49.99 Hz. A one-second simulation step is used, so $f_h[s]$ in Equation (5) is the power grid frequency for the current second, which corresponds to *sec* in Figure 3. When the frequency is under 49.9 Hz, the full power capacity of the bid, negative of *capacity(a)*, is the discharge power.

$$ctrl(a) = \frac{capacity(a)}{49.99\ \text{Hz} - 49.9\ \text{Hz}} \times f_h[s] - \frac{49.99\ \text{Hz} \times capacity(a)}{49.99\ \text{Hz} - 49.9\ \text{Hz}}, \quad (5)$$

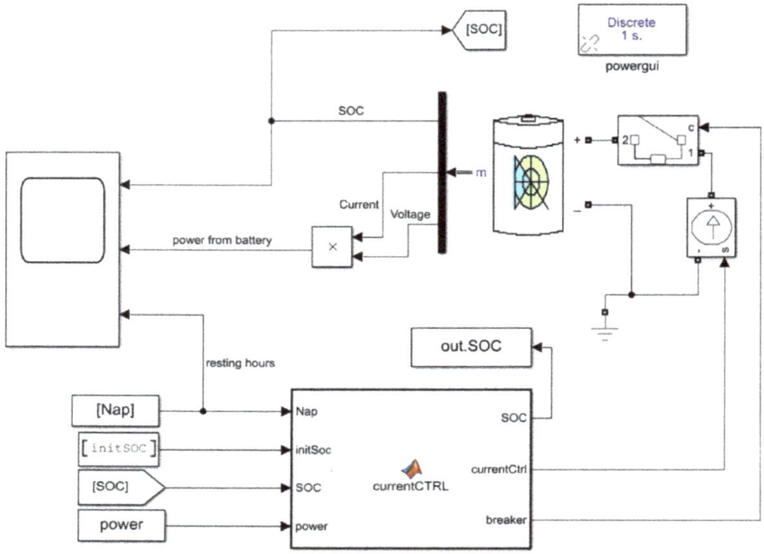


**Figure 4.** Battery simulation model.

The case of frequency in the range 50.01–50.1 Hz is symmetric and is described by Equation (6).

$$ctrl(a) = \frac{capacity(a)}{50.1\ \text{Hz} - 50.01\ \text{Hz}} \times f_h[s] - \frac{50.01\ \text{Hz} \times capacity(a)}{50.1\ \text{Hz} - 50.01\ \text{Hz}}, \quad (6)$$

*ctrl(a)* from Equations (5) and (6) is the 'power' input in Figure 4. Figure 5 shows how 'power' is computed from the frequency according to a software implementation of Equations (5) and (6). In this example, *a* = 0, so maximum *capacity(a)* is 1 MW. The slight differences between the red and blue curves are due to the deadband, e.g., according to Equation (5), 'power' is 0 when the frequency is 49.99 Hz. When frequency exceeds 50.1 Hz Equation (6) no longer applies, and the 'power' remains at 1 MW.

The battery in Figure 4 is an instance of the 'Battery' from Simulink's Simscape library [51]. The charging and discharging losses of the battery simulation component are according to the equations for the lithium-ion battery type in [51]. The battery has been parameterized as specified in Table 3. The *OoB* limits for the function *sim(pow)* in Table 2 are defined as 5% and 95% *SoC*.

The battery simulation model is an open-loop system, where the battery's behavior is controlled with a controlled current source. The controlled current source receives its control signal from the 'CurrentCTRL' (see Figure 4) MATLAB function. Its main purpose is to convert the 'power' input to a current signal for the controlled current source. This is done by dividing the signal by the battery's 'nominal voltage' (Table 3) when there is a bid for that hour. Otherwise, the battery rests, which is indicated by the 'Nap' input to 'CurrentCTRL'. During rest hour, the battery will charge or discharge towards *SOC* 50% with constant current. The charging and discharging are configured so that the *SoC* has time to reach 50% by the end of the rest hour, regardless of the initial *SoC*. 'CurrentCTRL' also keeps track of the $SoC$ during the simulation. The $SoC$ vector is passed to the MATLAB function that implements the $bounds(OoB_s)$ function in Table 2.

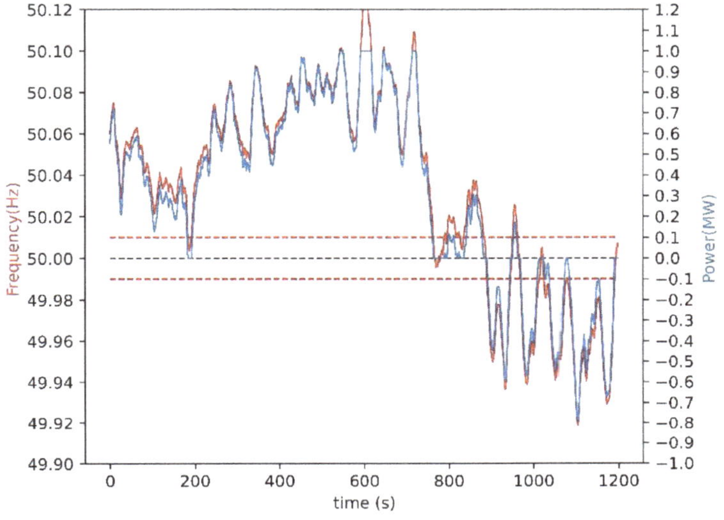


**Figure 5.** Power as a function of frequency (Equations (5) and (6)) for the time period 4 September 2020 00:32:59–00:52:59. Red horizontal lines are the deadband limits. Black horizontal line is the nominal frequency and the battery idle state.

**Table 3.** Parameters of the battery in Simulink.

| Parameter | Value |
|---|---|
| Type | Lithium-ion |
| Nominal voltage | 1200 V |
| Rated capacity | 1400 Ah |
| Battery response time | 0.1 s |
| Simulate temperature effects | No |
| Simulate aging effects | No |
| Discharge parameters: determine from the nominal parameters of the battery | Yes |

The battery simulation model was wrapped in custom Python code that implements an interface similar to the environments in the OpenAI Gym collection [52], which has been used in several recent publications on RL applications in the energy domain [53–58]. This interface defines the functions $reset(S[0])$ and $step(a)$. $reset(S[0])$ is called in Figure 3 at the beginning of each day and assigns a random value to the $SoC$, which ensures that the RL can continue to gain new experiences when the same day is used several times in the training phase. In our implementation, the $SoC$ is assigned a random value from a continuous uniform distribution with bounds 35% and 65%. The shaded area in Figure 3 is the $step(a)$ function, which receives the action from the RL agent and returns the reward and the next state.

### 4.2. Bidding Agent

The RL agent is implemented as a densely connected neural network. Its hyperparameters were determined experimentally and are presented in Table 4. The input layer has two nodes, since the state vector $S[h]$ has two elements. The output layer has four nodes, one for each possible value of the action $a$. An epsilon greedy exploration strategy is used, so the probability of selecting a random action is initially 1 and is decreased by the epsilon decay factor at the end of each day in the algorithm of Figure 3. The algorithm in Figure 3 collects all 24 experiences gained over one day into an experience trace, which is used to train the neural network, and a discount of 0.5 is applied to the trace.

**Table 4.** Hyperparameters of the neural network.

| Hyperparameter | Value |
|---|---|
| Number of hidden layers | 1 |
| Number of nodes in input layer | 2 |
| Number of nodes in hidden layer | 20 |
| Number of nodes in output layer | 4 |
| Epsilon decay | 0.998 |
| Learning rate | 0.01 |
| Discount factor | 0.5 |
| Hidden layer activation function | Sigmoid |
| Output layer activation function | Softmax |
| Dropout | Not used |
| Optimizer | Adam |

Further work is possible for the optimization of hyperparameters. Automated machine learning methods for neural architecture search can identify the optimal set of layers for a deep neural network. Once the architecture has been fixed, automatic hyperparameter tuning methods can optimize the remaining hyperparameters. However, these techniques are in general not directly applicable to reinforcement learning [59]. Recent applications to deep reinforcement learning are a promising approach for improving our neural network architecture and hyperparameters [60].

The time range of 1 September 2020–31 October 2020 was used for training and validation. A value of 110 was used for $reputation_{factor}$. Out of these 61 days, 11 randomly selected days were used for validation and the rest were used for training. A random seed was defined to ensure the repeatability of the results. Figure 6 shows some insights into the training process after 2, 4, 6, 8 and 10 epochs.

On the left of Figure 6, the actions selected by the trained RL agent are shown for each state (the state is defined by the combination of $R$ on the vertical axis and $FCR\text{-}N_{fcast}$ on the horizontal axis). Analyzing Equations (2)–(4), it can be seen that the positive component of the reward is directly proportional to the price of the FCR-N market, which is approximated by $FCR\text{-}N_{fcast}$. However, the negative component of the reward is only partially proportional to the price. Thus, at higher prices, the benefits should outweigh the penalties, so it is expected that the agent will learn to prefer resting during low-price hours. Accordingly, in Figure 6, it is observed that resting actions concentrate on the left of the price forecast axis. With respect to the vertical axis, it is expected that the likelihood of penalties increases when the battery has operated for several hours without resting, so it is expected that the agent will learn to prefer resting on the lower part of the vertical axis. The combined effect of these two learning outcomes is that the best states for resting are in the bottom-left corner and the best states for bidding are in the top-right corner. By observing the progression of the left-hand charts in Figure 6, it is evident that the agent has learned this behavior.

On the right of Figure 6, the bidding actions taken by the agent are shown for one of the validation days, 4 September 2020. After epoch 2 (Figure 6a), the chart on the left shows that the agent has learned to use three actions: rest, bid 600 kW and bid 800 kW. Only the rest and 800 kW actions are used in the chart on the left (the blue bars show the bid size with 0 meaning rest). The red prices are the forecasted market price. As training progresses over subsequent epochs 4, 6, 8 and 10 (Figure 6b-e), the agent learns to use only two actions: resting and 800 kW bid. The agent also learns to schedule the rest actions for hours with low price. The figure does not show penalties and rewards, which are discussed next in Section 5. The purpose of the discussion in this section was to give insights into the RL training process and the behavior learned by the RL agent.

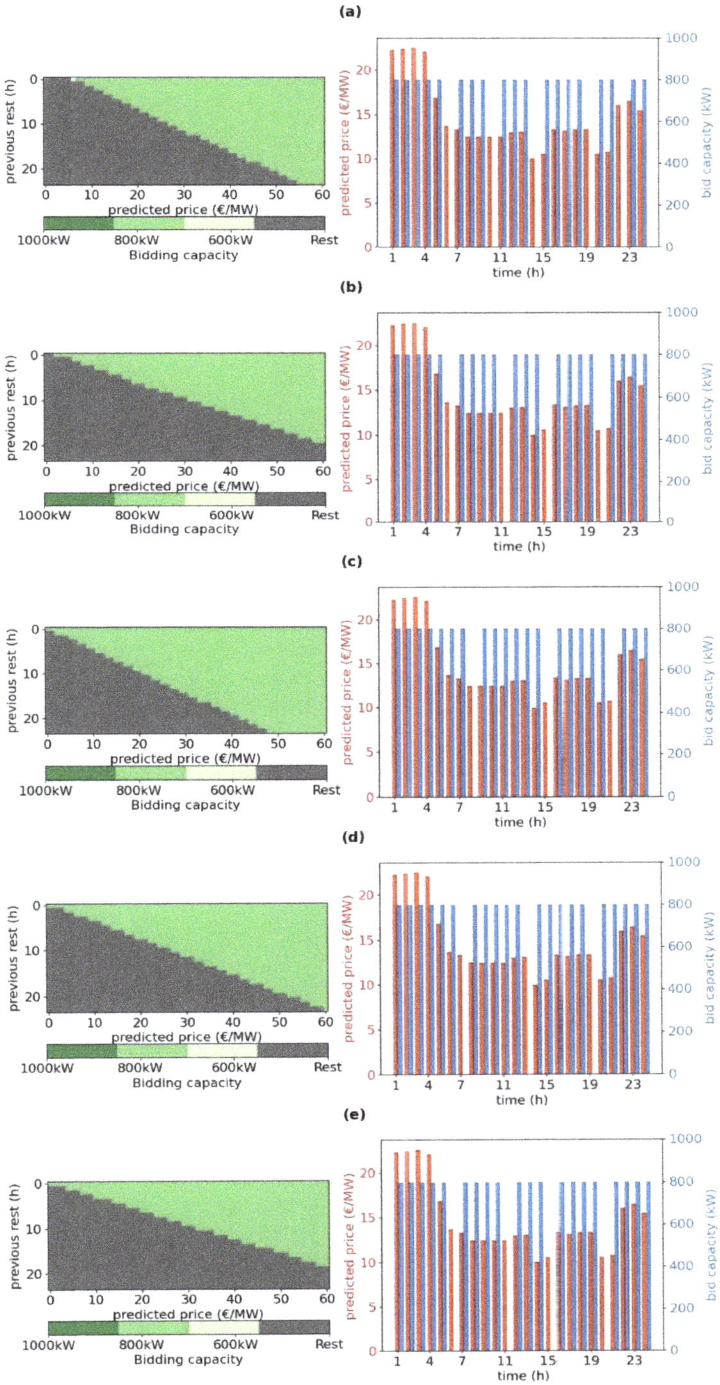


**Figure 6.** The actions taken by the trained reinforcement learning agent (left) and the resulting bidding behavior on 4 September 2020 (right) after epochs 2 (**a**), 4 (**b**), 6 (**c**), 8 (**d**) and 10 (**e**).

## 5. Result

Figure 7 shows the cumulative reward for all of the days in the training set, and Figure 8 shows the cumulative reward for the days in the validation set. The results for the training and validation sets start to stabilize after 20 epochs, so training was stopped after 35 epochs.

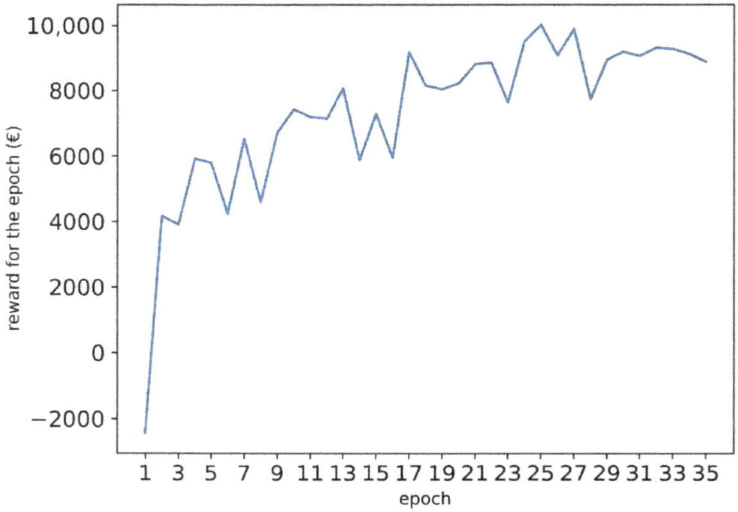


**Figure 7.** The cumulative reward for all of the training days.

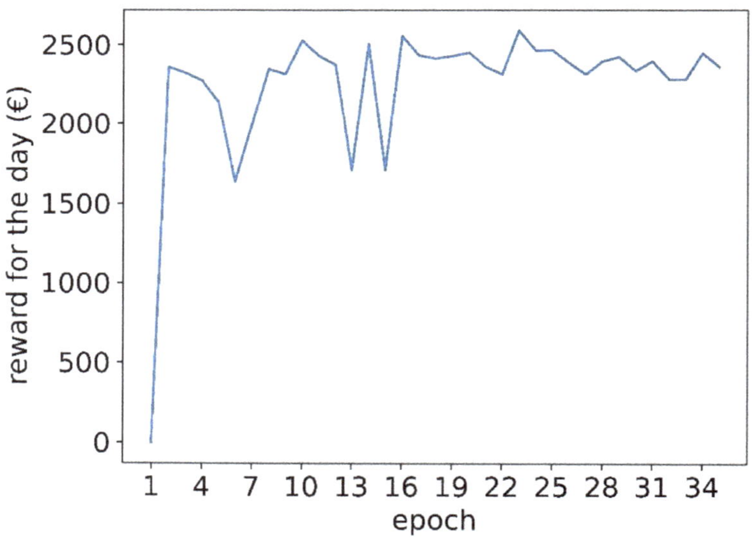


**Figure 8.** The cumulative reward for all the validation days.

Figure 9 shows the cumulative reward for one of the validation days, 4 September 2020. This is the sum of the rewards for each hour of the day. At nine epochs, there is a relatively low reward. This epoch is analyzed further in Figure 10. The chart in Figure 10a is similar to the charts on the right in Figure 6. The chart in Figure 10b is the reward and the actual market price. The chart in Figure 10c is the penalties. It is observed that due to only three resting hours for the entire day, the battery fails to provide the reserve capacity and incurs significant penalties on hour 17, which explains the low reward for epoch 9 in Figure 9. Figure 11 shows a similar chart after 35 epochs of training. The chart in Figure 11a shows that the agent has learned to rest more frequently, and generally, the rest occurs when there is a low price forecast. Although the increased resting reduces the market revenues (Figure 11b), there are no penalties (Figure 11c). Thus, the agent at 35 epochs plays safer than the agent at 10 epochs, resulting in a fairly good reward at 35 epochs, although the reward is not as high as in some of the earlier epochs, when the agent was resting less and thus making riskier bids. The risks are due to the unpredictable need to charge or discharge the battery when participating in PFR. The need depends on the occurrence of grid frequency deviations. There is a lack of research for predicting such deviations day-ahead (which is when the PFR bids must be placed), so our agent does not have information to learn the likelihood of charging or discharging needs for any particular hour. However, the results show that based on the available market forecasts, the agent learns to bid intelligently under uncertainty, balancing revenues and risk of penalties.

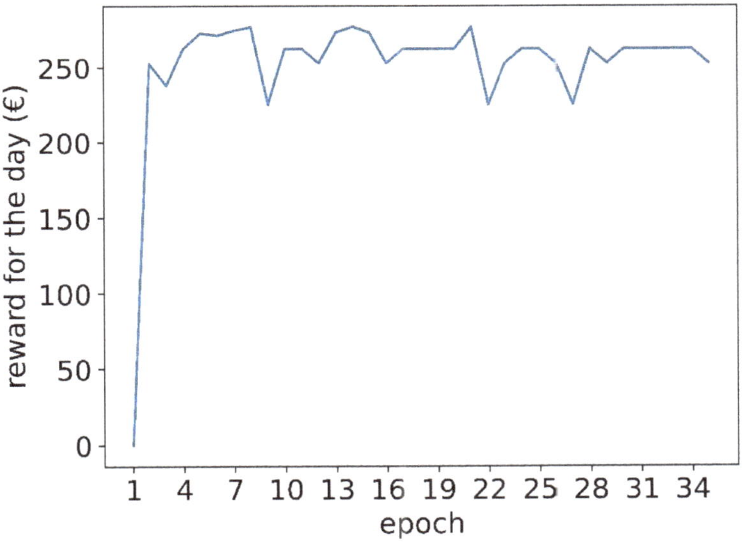


**Figure 9.** The cumulative reward for all of the hours for one of the validation days (4 September 2020).

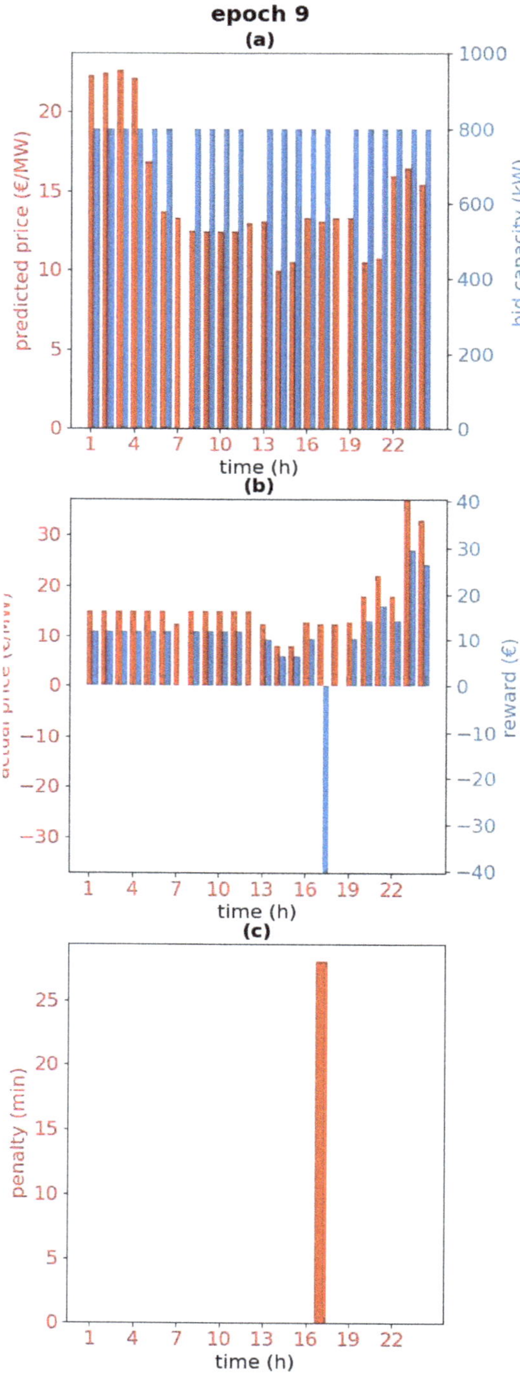


**Figure 10.** Bids and predicted price (**a**), reward and actual market price (**b**) and penalties (**c**) for the validation day 4 September 2020 at 9 epochs.

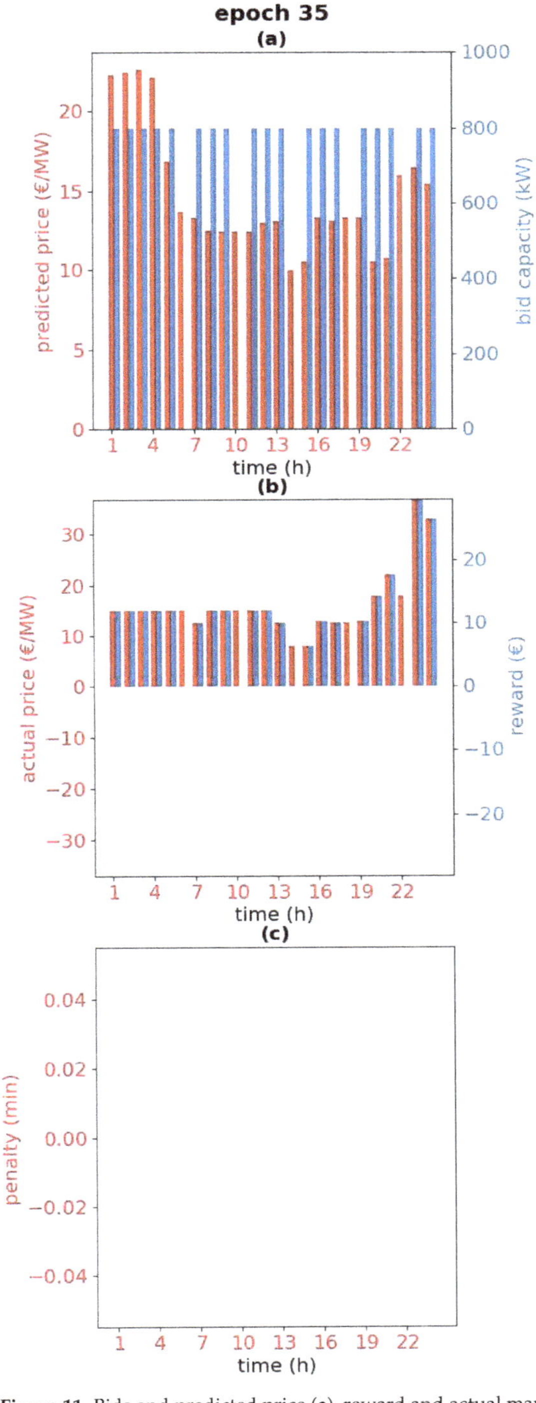


**Figure 11.** Bids and predicted price (**a**), reward and actual market price (**b**) and penalties (**c**) for the validation day 4 September 2020 at 35 epochs.

The results shown in this paper have been obtained by using a value of 110 for the $reputation_{factor}$. Figure 12 shows how the results would change if the value of $reputation_{factor}$ is varied. The experiment described in this paper was repeated for the following values: 10, 30, 50, 70, 90, 110, 130, 150, 170 and 190. Each repetition of the experiment resulted in one dot in the figure, labeled with the value of $reputation_{factor}$. According to Equations (3) and (4), a higher value of $reputation_{factor}$ will result in a large negative component in the reward whenever the battery is unavailable. The duration of this unavailability is $penalty_{min}$ on the x-axis. The compensation on the y-axis is according to Equation (1). As $reputation_{factor}$ is increased, it is expected that the RL agent learns to be more careful in avoiding penalties, either by resting more or by bidding a lower capacity, thus reducing the likelihood of the battery being unavailable. The result of this should be decreasing compensation and decreasing penalties as $reputation_{factor}$ increases. This trend is visible in Figure 12. The dots for 10, 30 and 50 are very close to each other and overlap in the figure. This is because the compensation outweighs the penalties, so the agent learns to ignore the penalties and only tries to maximize the compensation. At a value of 70, the penalties are drastically reduced, without a loss of compensation. In fact, the compensation is slightly higher, which can be understood from Equation (1): there is no compensation for the minutes during which the battery is unavailable. As $reputation_{factor}$ is increased to 90 and beyond, the trend that was mentioned above is observed: the agent bids more carefully resulting in a slight decrease in compensation as well as in the penalties. Looking at the relative vertical positions of the dots, 130 is an outlier in this trend. Further, 110 and 150 do not fully fit into the trend. The validation set is 11 days, so a longer set would be expected to result in a clearer trend. From the results, it is concluded that it is advantageous to use a $reputation_{factor}$ of at least 70. The use of a higher value is a business decision, depending on whether a decrease in compensation is considered desirable in order to decrease the penalties. As has been explained in the context of Equation (3), the potential business impact of incurring excessive penalties is very severe, but the market operator does not publish any quantitative criteria for what it considers to be excessive penalties, so for that reason, the choice of value for $reputation_{factor}$ is left as a business decision.

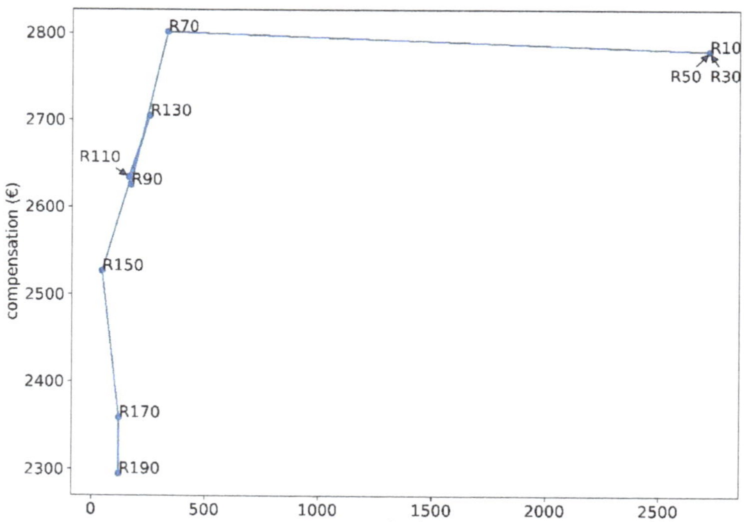


**Figure 12.** Compensation versus $penalty_{min}$. The dots show the result of running the experiment with different values of $reputation_{factor}$. Each dot is labeled with the corresponding value of $reputation_{factor}$.

## 6. Conclusions

In the literature review, a research gap was identified for RL-based energy management solutions that take into account market participation and cope with real-time requirements for the energy resources that participate in the markets. In this paper, PFR was selected as an application in which revenues depend on battery capacity that is bid on hourly markets, as well as penalties that occur on the timeframe of seconds if the battery is unavailable due to its *SoC* being OoB. The problem formulation addressed the realities of an online deployment, in which bids on the PFR market must be done on the day before, when it is not possible to accurately predict the *SoC*, as the requirement to discharge and charge the battery on a PFR market is dependent on power grid frequency deviations, which cannot be accurately predicted day-ahead. Thus, the state information for the RL was limited to information that is available at bidding time. It was observed that the agent learned behavior that took into account the benefits of bidding on high-price hours and the increased risk of penalties of participating in PFR markets for several hours without allowing the battery to rest.

A novel methodology integrating a real-time battery simulation with a reinforcement learning agent bidding on hourly markets was proposed in this article. The main finding is that this approach promises to achieve the dual goal of maximizing market revenues while minimizing penalties caused by short-term failures to provide the frequency reserve. In further work, the reliability of the methodology can be improved by addressing the following limitations: Firstly, a 2-month dataset was used, so market and grid frequency data for a longer time-period can be collected and preprocessed. Secondly, the state space can be broadened with any variables that may have an impact on the power grid frequency. Although it is not possible to accurately predict the grid frequency in a day-ahead bidding scenario, some feature engineering based on historical frequency data is an avenue of further research. Finally, automated machine learning methods that have recently emerged for reinforcement learning applications can be used to search for the optimal neural network architecture and hyperparameters.

For further work, batteries for supporting photovoltaic installations in residential and commercial buildings are an application area that would benefit from optimization on the two timescales that have been considered in this paper. Maximum power point tracking (MPPT) algorithms have been proposed to control the battery and thus create an ideal load for photovoltaic generation. However, such batteries have other uses related to shifting power consumption from the grid and possible photovoltaic power sales to the grid, taking into account variable electricity prices. The MPPT and variable electricity price exploitation are two optimization tasks that occur on two different timescales but cannot be addressed separately, since they both affect the *SoC* of the same battery.

**Author Contributions:** Conceptualization, H.A. and S.S.; methodology, H.A. and S.S.; software, H.A. and R.S.; validation, H.A., R.S., S.S. and V.V.; investigation, H.A., R.S. and S.S.; resources, H.A., R.S., S.S. and V.V.; data curation, H.A.; writing—original draft preparation, H.A. and S.S.; visualization, H.A. and R.S.; supervision, S.S. and V.V.; project administration, S.S. and V.V.; funding acquisition, S.S. All authors have read and agreed to the published version of the manuscript.

**Funding:** This research was supported by Business Finland grant 7439/31/2018.

**Institutional Review Board Statement:** Not applicable.

**Informed Consent Statement:** Not applicable.

**Data Availability Statement:** Data are contained within the article.

**Acknowledgments:** The authors thank Kalle Rantala for technical support with high-performance computing infrastructure.

**Conflicts of Interest:** The authors declare no conflict of interest.

## References


1. Peters, I.M.; Breyer, C.; Jaffer, S.A.; Kurtz, S.; Reindl, T.; Sinton, R.; Vetter, M. The role of batteries in meeting the PV terawatt challenge. *Joule* **2021**, *5*, 1353–1370. [CrossRef]
2. de Siqueira Silva, L.M.; Peng, W. Control strategy to smooth wind power output using battery energy storage system: A review. *J. Energy Storage* **2021**, *35*, 102252. [CrossRef]
3. Sepúlveda-Mora, S.B.; Hegedus, S. Making the case for time-of-use electric rates to boost the value of battery storage in commercial buildings with grid connected PV systems. *Energy* **2021**, *218*, 119447. [CrossRef]
4. Loukatou, A.; Johnson, P.; Howell, S.; Duck, P. Optimal valuation of wind energy projects co-located with battery storage. *Appl. Energy* **2021**, *283*, 116247. [CrossRef]
5. Akagi, S.; Yoshizawa, S.; Ito, M.; Fujimoto, Y.; Miyazaki, T.; Hayashi, Y.; Tawa, K.; Hisada, T.; Yano, T. Multipurpose control and planning method for battery energy storage systems in distribution network with photovoltaic plant. *Int. J. Electr. Power Energy Syst.* **2020**, *116*, 105485. [CrossRef]
6. Nefedov, E.; Sierla, S.; Vyatkin, V. Internet of energy approach for sustainable use of electric vehicles as energy storage of prosumer buildings. *Energies* **2018**, *11*, 2165. [CrossRef]
7. Ge, X.; Ahmed, F.W.; Rezvani, A.; Aljojo, N.; Samad, S.; Foong, L.K. Implementation of a novel hybrid BAT-Fuzzy controller based MPPT for grid-connected PV-battery system. *Control. Eng. Pract.* **2020**, *98*, 104380. [CrossRef]
8. Aldosary, A.; Ali, Z.M.; Alhaider, M.M.; Ghahremani, M.; Dadfar, S.; Suzuki, K. A modified shuffled frog algorithm to improve MPPT controller in PV System with storage batteries under variable atmospheric conditions. *Control. Eng. Pract.* **2021**, *112*, 104831. [CrossRef]
9. Ciupageanu, D.; Barelli, L.; Lazaroiu, G. Real-time stochastic power management strategies in hybrid renewable energy systems: A review of key applications and perspectives. *Electr. Power Syst. Res.* **2020**, *187*, 106497. [CrossRef]
10. Lin, L.; Jia, Y.; Ma, M.; Jin, X.; Zhu, L.; Luo, H. Long-term stable operation control method of dual-battery energy storage system for smoothing wind power fluctuations. *Int. J. Electr. Power Energy Syst.* **2021**, *129*, 106878. [CrossRef]
11. Ryu, A.; Ishii, H.; Hayashi, Y. Battery smoothing control for photovoltaic system using short-term forecast with total sky images. *Electr. Power Syst. Res.* **2021**, *190*, 106645. [CrossRef]
12. Subramanya, R.; Yli-Ojanperä, M.; Sierla, S.; Hölttä, T.; Valtakari, J.; Vyatkin, V. A virtual power plant solution for aggregating photovoltaic systems and other distributed energy resources for northern european primary frequency reserves. *Energies* **2021**, *14*, 1242. [CrossRef]
13. Koller, M.; Borsche, T.; Ulbig, A.; Andersson, G. Review of grid applications with the Zurich 1MW battery energy storage system. *Electr. Power Syst. Res.* **2015**, *120*, 128–135. [CrossRef]
14. Giovanelli, C.; Sierla, S.; Ichise, R.; Vyatkin, V. Exploiting artificial neural networks for the prediction of ancillary energy market prices. *Energies* **2018**, *11*, 1906. [CrossRef]
15. Lund, H.; Hvelplund, F.; Østergaard, P.A.; Möller, B.; Mathiesen, B.V.; Karnøe, P.; Andersen, A.N.; Morthorst, P.E.; Karlsson, K.; Münster, M.; et al. System and market integration of wind power in Denmark. *Energy Strategy Rev.* **2013**, *1*, 143–156. [CrossRef]
16. Bialek, J. What does the GB power outage on 9 August 2019 tell us about the current state of decarbonised power systems? *Energy Policy* **2020**, *146*, 111821. [CrossRef]
17. Papadogiannis, K.A.; Hatziargyriou, N.D. Optimal allocation of primary reserve services in energy markets. *IEEE Trans. Power Syst.* **2004**, *19*, 652–659. [CrossRef]
18. Pavić, I.; Capuder, T.; Kuzle, I. Low carbon technologies as providers of operational flexibility in future power systems. *Appl. Energy* **2016**, *168*, 724–738. [CrossRef]
19. Zecchino, A.; Prostejovsky, A.M.; Ziras, C.; Marinelli, M. Large-scale provision of frequency control via V2G: The Bornholm power system case. *Electr. Power Syst. Res.* **2019**, *170*, 25–34. [CrossRef]
20. Malik, A.; Ravishankar, J. A hybrid control approach for regulating frequency through demand response. *Appl. Energy* **2018**, *210*, 1347–1362. [CrossRef]
21. Borsche, T.S.; de Santiago, J.; Andersson, G. Stochastic control of cooling appliances under disturbances for primary frequency reserves. *Sustain. Energy Grids Netw.* **2016**, *7*, 70–79. [CrossRef]
22. Herre, L.; Tomasini, F.; Paridari, K.; Söder, L.; Nordström, L. Simplified model of integrated paper mill for optimal bidding in energy and reserve markets. *Appl. Energy* **2020**, *279*, 115857. [CrossRef]
23. Ramírez, M.; Castellanos, R.; Calderón, G.; Malik, O. Placement and sizing of battery energy storage for primary frequency control in an isolated section of the Mexican power system. *Electr. Power Syst. Res.* **2018**, *160*, 142–150. [CrossRef]
24. Killer, M.; Farrokhseresht, M.; Paterakis, N.G. Implementation of large-scale li-ion battery energy storage systems within the EMEA region. *Appl. Energy* **2020**, *260*, 114166. [CrossRef]
25. Oudalov, A.; Chartouni, D.; Ohler, C. Optimizing a battery energy storage system for primary frequency control. *IEEE Trans. Power Syst.* **2007**, *22*, 1259–1266. [CrossRef]
26. Andrenacci, N.; Pede, G.; Chiodo, E.; Lauria, D.; Mottola, F. Tools for life cycle estimation of energy storage system for primary frequency reserve. In Proceedings of the International Symposium on Power Electronics, Electrical Drives, Automation and Motion (SPEEDAM), Amalfi, Italy, 20–22 June 2018; pp. 1008–1013. [CrossRef]
27. Karbouj, H.; Rather, Z.H.; Flynn, D.; Qazi, H.W. Non-synchronous fast frequency reserves in renewable energy integrated power systems: A critical review. *Int. J. Electr. Power Energy Syst.* **2019**, *106*, 488–501. [CrossRef]

28. Srinivasan, L.; Markovic, U.; Vayá, M.G.; Hug, G. Provision of frequency control by a BESS in combination with flexible units. In Proceedings of the 5th IEEE International Energy Conference (ENERGYCON), Limassol, Cyprus, 3–7 June 2018; pp. 1–6. [CrossRef]
29. Phan, B.C.; Lai, Y. Control strategy of a hybrid renewable energy system based on reinforcement learning approach for an isolated microgrid. *Appl. Sci.* **2019**, *9*, 4001. [CrossRef]
30. Li, W.; Cui, H.; Nemeth, T.; Jansen, J.; Ünlübayir, C.; Wei, Z.; Zhang, L.; Wang, Z.; Ruan, J.; Dai, H.; et al. Deep reinforcement learning-based energy management of hybrid battery systems in electric vehicles. *J. Energy Storage* **2021**, *36*, 102355. [CrossRef]
31. Chen, Z.; Hu, H.; Wu, Y.; Xiao, R.; Shen, J.; Liu, Y. Energy management for a power-split plug-in hybrid electric vehicle based on reinforcement learning. *Appl. Sci.* **2018**, *8*, 2494. [CrossRef]
32. Sui, Y.; Song, S. A multi-agent reinforcement learning framework for lithium-ion battery scheduling problems. *Energies* **2020**, *13*, 1982. [CrossRef]
33. Muriithi, G.; Chowdhury, S. Optimal energy management of a grid-tied solar pv-battery microgrid: A reinforcement learning approach. *Energies* **2021**, *14*, 2700. [CrossRef]
34. Kim, S.; Lim, H. Reinforcement learning based energy management algorithm for smart energy buildings. *Energies* **2018**, *11*, 2010. [CrossRef]
35. Lee, S.; Choi, D. Reinforcement learning-based energy management of smart home with rooftop solar photovoltaic system, energy storage system, and home appliances. *Sensors* **2019**, *19*, 3937. [CrossRef]
36. Lee, S.; Choi, D. Energy management of smart home with home appliances, energy storage system and electric vehicle: A hierarchical deep reinforcement learning approach. *Sensors* **2020**, *20*, 2157. [CrossRef]
37. Roesch, M.; Linder, C.; Zimmermann, R.; Rudolf, A.; Hohmann, A.; Reinhart, G. Smart grid for industry using multi-agent reinforcement learning. *Appl. Sci.* **2020**, *10*, 6900. [CrossRef]
38. Kim, J.; Lee, B. Automatic P2P Energy trading model based on reinforcement learning using long short-term delayed reward. *Energies* **2020**, *13*, 5359. [CrossRef]
39. Wang, N.; Xu, W.; Shao, W.; Xu, Z. A q-cube framework of reinforcement learning algorithm for continuous double auction among microgrids. *Energies* **2019**, *12*, 2891. [CrossRef]
40. Mbuwir, B.V.; Ruelens, F.; Spiessens, F.; Deconinck, G. Battery energy management in a microgrid using batch reinforcement learning. *Energies* **2017**, *10*, 1846. [CrossRef]
41. Zsembinszki, G.; Fernández, C.; Vérez, D.; Cabeza, L.F. Deep Learning optimal control for a complex hybrid energy storage system. *Buildings* **2021**, *11*, 194. [CrossRef]
42. Lee, H.; Ji, D.; Cho, D. Optimal design of wireless charging electric bus system based on reinforcement learning. *Energies* **2019**, *12*, 1229. [CrossRef]
43. Oh, E. Reinforcement-learning-based virtual energy storage system operation strategy for wind power forecast uncertainty management. *Appl. Sci.* **2020**, *10*, 6420. [CrossRef]
44. Tsianikas, S.; Yousefi, N.; Zhou, J.; Rodgers, M.D.; Coit, D. A storage expansion planning framework using reinforcement learning and simulation-based optimization. *Appl. Energy* **2021**, *290*, 116778. [CrossRef]
45. Sidorov, D.; Panasetsky, D.; Tomin, N.; Karamov, D.; Zhukov, A.; Muftahov, I.; Dreglea, A.; Liu, F.; Li, Y. Toward zero-emission hybrid AC/DC power systems with renewable energy sources and storages: A case study from Lake Baikal region. *Energies* **2020**, *13*, 1226. [CrossRef]
46. Xu, B.; Shi, J.; Li, S.; Li, H.; Wang, Z. Energy consumption and battery aging minimization using a q-learning strategy for a battery/ultracapacitor electric vehicle. *Energy* **2021**, *229*, 120705. [CrossRef]
47. Zhang, G.; Hu, W.; Cao, D.; Liu, W.; Huang, R.; Huang, Q.; Chen, Z.; Blaabjerg, F. Data-driven optimal energy management for a wind-solar-diesel-battery-reverse osmosis hybrid energy system using a deep reinforcement learning approach. *Energy Convers. Manag.* **2021**, *227*, 113608. [CrossRef]
48. Fingrid. The Technical Requirements and the Prequalification Process of Frequency Containment Reserves (FCR). Available online: https://www.fingrid.fi/globalassets/dokumentit/en/electricity-market/reserves/appendix3---technical-requirements-and-prequalification-process-of-fcr.pdf (accessed on 6 July 2021).
49. Fingrid. Fingridin reservikaupankäynti ja tiedonvaihto -ohje. Available online: https://www.fingrid.fi/globalassets/dokumentit/fi/sahkomarkkinat/reservit/fingridin-reservikaupankaynti-ja-tiedonvaihto--ohje.pdf (accessed on 6 July 2021).
50. Fingrid. Ehdot ja edellytykset taajuudenvakautusreservin (FCR) toimittajalle. Available online: https://www.fingrid.fi/globalassets/dokumentit/fi/sahkomarkkinat/reservit/fcr-liite1---ehdot-ja-edellytykset.pdf (accessed on 6 July 2021).
51. MathWorks. Battery—Generic Battery Model. Available online: https://se.mathworks.com/help/physmod/sps/powersys/ref/battery.html (accessed on 6 July 2021).
52. Brockman, G.; Cheung, V.; Pettersson, L.; Schneider, J.; Schulman, J.; Tang, J.; Zaremba, W. Openai gym. *arXiv* **2016**, arXiv:1606.01540.
53. Avila, L.; De Paula, M.; Trimboli, M.; Carlucho, I. Deep reinforcement learning approach for MPPT control of partially shaded PV systems in Smart Grids. *Appl. Soft Comput.* **2020**, *97*, 106711. [CrossRef]
54. Zhang, Z.; Chong, A.; Pan, Y.; Zhang, C.; Lam, K.P. Whole building energy model for HVAC optimal control: A practical framework based on deep reinforcement learning. *Energy Build.* **2019**, *199*, 472–490. [CrossRef]

55. Azuatalam, D.; Lee, W.; de Nijs, F.; Liebman, A. Reinforcement learning for whole-building HVAC control and demand response *Energy AI* **2020**, *2*, 100020. [CrossRef]
56. Brandi, S.; Piscitelli, M.S.; Martellacci, M.; Capozzoli, A. Deep reinforcement learning to optimise indoor temperature control and heating energy consumption in buildings. *Energy Build.* **2020**, *224*, 110225. [CrossRef]
57. Nakabi, T.A.; Toivanen, P. Deep reinforcement learning for energy management in a microgrid with flexible demand. *Sustain. Energy Grids Netw.* **2021**, *25*, 100413. [CrossRef]
58. Schreiber, T.; Eschweiler, S.; Baranski, M.; Müller, D. Application of two promising reinforcement learning algorithms for load shifting in a cooling supply system. *Energy Build.* **2020**, *229*, 110490. [CrossRef]
59. He, X.; Zhao, K.; Chu, X. AutoML: A survey of the state-of-the-art. *Knowl. Based Syst.* **2021**, *212*, 106622. [CrossRef]
60. Franke, J.K.; Köhler, G.; Biedenkapp, A.; Hutter, F. Sample-efficient automated deep reinforcement learning. *arXiv* **2020** arXiv:2009.01555.

energies

Article

# Decentralized Prosumer-Centric P2P Electricity Market Coordination with Grid Security

**Duarte Kazacos Winter, Rahul Khatri and Michael Schmidt ***

Institute of Energy Systems Technology, Offenburg University of Applied Sciences, 77652 Offenburg, Germany; duarte.kazacos@outlook.com (D.K.W.); rahul-khatri@outlook.com (R.K.)
* Correspondence: schmidt@hs-offenburg.de

**Abstract:** The increasing number of prosumers and the accompanying greater use of decentralised energy resources (DERs) bring new opportunities and challenges for the traditional electricity systems and the electricity markets. Microgrids, virtual power plants (VPPs), peer-to-peer (P2P) trading and federated power plants (FPPs) propose different schemes for prosumer coordination and have the potential of becoming the new paradigm of electricity market and power system operation. This paper proposes a P2P trading scheme for energy communities that negotiates power flows between participating prosumers with insufficient renewable power supply and prosumers with surplus supply in such a way that the community welfare is maximized while avoiding critical grid conditions. For this purpose, the proposed scheme is based on an Optimal Power Flow (OPF) problem with a Multi-Bilateral Economic Dispatch (MBED) formulation as an objective function. The solution is realized in a fully decentralized manner on the basis of the Relaxed Consensus + Innovations (RCI) algorithm. Network security is ensured by a tariff-based system organized by a network agent that makes use of product differentiation capabilities of the RCI algorithm. It is found that the proposed mechanism accurately finds and prevents hazardous network operations, such as over-voltage in grid buses, while successfully providing economic value to prosumers' renewable generation within the scope of a P2P, free market.

**Keywords:** renewable energy; peer-to-peer; electricity market, economic dispatch; consensus + innovations; distributed energy resources



**Citation:** Kazacos Winter, D.; Khatri, R.; Schmidt, M. Decentralized Prosumer-Centric P2P Electricity Market Coordination with Grid Security. *Energies* **2021**, *14*, 4665. https://doi.org/10.3390/en14154665

Academic Editors: Seppo Sierla and Abu-Siada Ahmed

Received: 21 June 2021
Accepted: 29 July 2021
Published: 1 August 2021



## 1. Introduction

Advances in technology for electricity generation, storage and smart meters as well as the declines in cost of access to these technologies pose a paradigm shift in the electricity sector as the number of consumers who are becoming active participants in the electricity market is rapidly increasing [1]. This opens the door for possibilities on how to envision the new electricity market design while facing the energy trilemma: Transitioning to zero-carbon emissions in energy generation while providing universal and secure access to energy and granting that access in an affordable manner [2]. One key element of such market design involving prosumers is the inclusion of a central coordination for energy management purposes that aggregates the demand response potential of a community of prosumers as suggested by authors in [3].

Individually, prosumers have little impact at a transmission level and the complexity of communication and transactional costs may outweigh the potential benefits of prosumers directly participating in the wholesale market [4]. In this scenario, if prosumers do not see benefits in being part of the retail market, they will see an incentive in going off-grid [5]. As prosumers migrate offline, the costs of the network will be shared among the rest of participants, increasing the value of the service provided by the network [6]. Peer-to-Peer (P2P) coordination comes with the benefits of building sufficient aggregated Distributed Energy Resources (DERs) as to influence the wholesale market and to extract value from trading at a prosumer level. P2P energy trading allows prosumers to directly sell and buy electricity to one another without the need for a central coordinator. However, in the shift toward a more decentralised energy supply, there is also an increasing demand for prosumers to play a greater role in ensuring grid stability, thus, coordination mechanisms should also include this functionality.

This paper, therefore, proposes a P2P coordination for energy communities that maximises the community welfare while avoiding critical grid conditions. It is especially intended for prosumers with their own renewable generation and the possibility of intermediate energy storage, who want to cover temporary energy shortages at low cost and to share surpluses profitably as part of an energy community. The community approach for peer-to-peer power exchange is modeled in form of a Multi-Bilateral Economic Dispatch (MBED) optimization problem [7].

In order to take into account physical constraints of the network such as voltage or thermal loading of network elements, this approach could now be combined with an optimal power flow (OPF) problem. P2P approaches including the network security as constraints for the OPF problem are, for instance, investigated in [8–13]. This paper takes a different approach by rather building the network requirements directly into the negotiation mechanism of the economic dispatch: Sorin et al. show in [7] how the the Relaxed Consensus + Innovations (RCI) method can be used to solve the MBED problem in a fully decentralized manner by means of agents. They also show how to implement a product differentiation into the negotiation scheme that can take additional trade characteristics of the exchanged power into account. The P2P scheme presented in this paper makes use of this product differentiation in order to iteratively increase the price penalty of harmful power exchanges until they are prevented. This approach renders the load flow and the underlying network topology itself to not directly be part of the optimization problem. In each iteration, only a normal load flow calculation has to be performed to provide input parameters for the price penalty terms. This calculation could be realized by a single designated agent controlled by the grid operator or community coordinator that then shares price signals with the peers.

Key aspects are found throughout this work. On the one hand, the possibility to establish an electricity market where participating agents can express preference and freely offer and demand power from sources at will, given the fact that the market mechanism is formed in a renewable energy-only local distribution grid of prosumers. On the other hand, the inclusion of a somewhat central grid operator at a local grid level has the rights to penalize or reward certain trades or to establish a P2P market where participating agents can freely express preference. Such agent is modeled with limited rights. Namely, its role is restricted to check for congestion problems at certain components of the network and to apply tariffs to the participating prosumers whose trades are found to cause those congestions.

The remaining part of this paper is structured as follows. Section 2 describes the methodology of the proposed P2P scheme which is based on a iterative process combining P2P negotiations and load flow calculations. For this, the modeling of the prosumers and their control strategies, the formulation of the P2P power exchange, the MBED formulation of the addressed optimization problem, the RCI method and the consideration of network constraints within the negotiation scheme are presented. Section 3 shows numerical results of an implementation of the proposed P2P scheme for a 12-bus-meshed network of prosumers. The agent-based P2P mechanism is realized in python, the network topology is built with the python open-source library pandapower [14], which is also used for load flow calculations. The results show the successful prevention of network problems and provides insight into the conversion behavior of the iterative trading scheme. Finally, in Section 4 key findings and future work are discussed.

## 2. Methodology

### 2.1. Prosumer Simulation

A prosumer it is thought of as wanting to maximize the consumption of its own energy production. For that, photovoltaic (PV) generation is consumed first. Subsequently, the battery energy storage system (BESS) is charged until a desired state of charge (SOC) is reached. Finally, the prosumer feeds the surplus generation into the community grid, should there be any. Similarly, a prosumer's default behavior at the time of no PV generation is to demand power from its BESS and then to buy from the grid as the BESS's SOC falls below a desired limit. Figure 1 shows a conceptual diagram of the basic prosumer components and the allowed power flow.

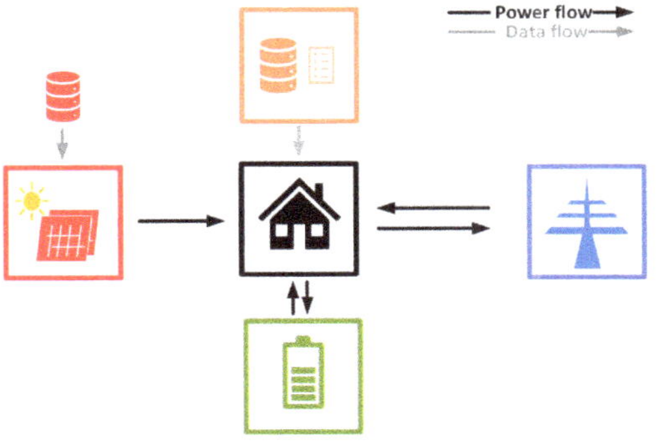


**Figure 1.** Representation of the power and data flow at a prosumer.

The black arrows denote the power flow whereas the grey arrows denote data flow. Data are here referred to the historical data of PV generation and a load profile in kWh for a given time step length, Δt. Naturally, the power flows in both directions for the exchange with both the BESS and the grid. That is, it does not account for any physical phenomena with regards to its internal components and, thus, it does not account for non-linear behavior of the SOC. The power balance equation that governs the prosumer's interaction with the power system is stated as follows:

$$P_{PV} + P_{BESS} + P_{Load} + P_{Grid} = 0 \quad (1)$$

where $P_{PV}$ is the power generated by the PV generator, $P_{BESS}$ is the power exchange with the BESS, $P_{Load}$ is the power requirements of the prosumer and $P_{Grid}$ is the power exchange with the grid point of connection.

### 2.2. Peer-to-Peer Market Coordination Model

An OPF calculation needs to account for congestion events in the power system and to access a solution in which they are tackled as effectively as possible. As such, to act and correct hazardous network operation by means of PV curtailment, load demand reduction or shifting and similar strategies by order of a central processing unit at a grid level through direct intervention might result in satisfactory grid management. However, the management of the OPF conditions by direct intervention and by controlling a grid coordination agent is not here believed to be a satisfactory enough solution for the prosumers involved. It is, instead, the intention of this work to provide an automatized solution that does not rely on such a direct intervention from a grid coordination agent on private prosumer installations and demand needs. It is believed that a free-market trading scheme provides a more satisfactory coordination than that of direct obligation, regulation and control.

A P2P market mechanism is applied in order to create an environment where prosumers can trade and exchange the power generation of their PV installations with other prosumers as agents of a market. This differs from a pool-based market, where all power generation is offered in a centralized exchange, where agents can then ask for the needed energy at the negotiated price. P2P markets are based on simultaneous negotiation of bilateral-trades where prosumers directly agree among each other on the energy and price traded. This offers a transparent clearing mechanism that involves all prosumers equally.

In this work, the role of a central coordinator is reduced to monitoring possible grid congestion occurrences. This market is built such that the valuation of certain trades is penalized in the case that the prosumer biding a power quantity is subsequently responsible for an over-voltage occurrence at its bus or thermal overload on the line through which its power is delivered. Therefore, the natural willingness of prosumers to trade among each other given the cost of certain trades is ultimately assuring an optimal network use. The proposed methodology allows for a simultaneous negotiation over the price and energy of multi-bilateral trades along a predefined trading scheme built directly in the negotiation mechanism itself. A full P2P market with a complete communication graph is considered where every agent, $n$, in the grid is able to trade with the rest of neighboring agents, $m$, on the local grid without the need to be physically connected to them. In order to model a full P2P trading scheme, the net power injection of any agent $n \in \Omega$ is considered as the sum of the bilaterally traded power quantities with neighboring agents $m \in \omega_n$. A neighbor here might not necessarily only be physically connected agents $m$ to agent $n$, but rather any other agent able to trade with the latter. A power quantity $P_n$ injected by agent $n$ can be defined as follows.

$$P_n = \sum_{m \in \omega_n} P_{nm} \tag{2}$$

The generator reference system is used, meaning that power injection takes positive values for those agents behaving as generators and negative values for consumers. The set of variables $\{P_{nm} | n \in \Omega, m \in \omega_n\}$ is the set of decision variables. An agent's limitation on the quantity of power to be injected is defined by the agent power boundaries that must satisfy the following.

$$\underline{P_n} \leq P_n \leq \overline{P_n} \tag{3}$$

A prosumer's boundary power injection is $\overline{P_n}, \underline{P_n} \geq 0$ at times when a prosumer behaves as a generator and $\overline{P_n}, \underline{P_n} \leq 0$, otherwise. Similarly, the decision variables assume the same criteria and are, therefore, $P_{nm} \geq 0$ for when prosumer $n$ is a generator and $P_{nm} \leq 0$, otherwise. The production cost or willingness to pay of a prosumer $n$ is modeled as in [15] by the quadratic function of the power set-point with positive parameters $a_n$, $b_n$ and $d_n$,

$$C_n(P_n) = \frac{1}{2} a_n P_n^2 + b_n P_n + d_n, \quad a_n, b_n, d_n \geq 0 \tag{4}$$

The cost of the generator or willingness to pay of the consumer is complemented with the cost of the bilaterally traded quantity towards each neighboring prosumer and has the following form:

$$\hat{C}_n(\boldsymbol{p})_n = \sum_{m \in \omega_n} c_{nm} P_{nm} \tag{5}$$

where $\boldsymbol{p} = (P_{nm})_{m \in \omega_n}$ is the vector of decision variables of prosumer $n$ and $c_{nm}$ is the bilateral trading coefficient imposed by prosumer $n$ on the trades with the rest of neighbors $m \in \omega_n$. The bilateral trading coefficient $c_{nm}$ is defined for the purpose of product differentiation. This coefficient encapsulates the information about what cost is applied to certain trades given the nature of the trading under the criterion $g \in G$ being $G$ the set of possible criteria of prosumer $n$ applied to its trades with prosumer $m$. The bilateral trading coefficient has the following form.

$$c_{nm} = \sum_{g \in G} c_n^g \gamma_{nm}^g \quad (6)$$

The bilateral trading coefficient indicates the objective value of a trade under the perspective of prosumer $n$ through the parameter $\gamma_{nm}^g \geq 0$, called the trade characteristic under criterion $g$. This allows for flexibility in the nature of the bilateral trading cost as to what value is given to what trading characteristic. For instance, the criteria $g$ could refer to the origin of the power traded, whether it comes from renewable sources or not or the distance between the producer and the consumer. The relative cost that prosumer $n$ applies on the trade is denoted by the parameter $c_n^g$, called the criterion value. The nature of the bilateral trading coefficient allows for expressing a variety of strategies for the conformation of a market mechanism. For instance, the trade criterion could be subjected to the decision of a centralized coordinator entity in order to express taxation or it could be set as a completely decentralized notion in order to express prosumer preference over certain trades.

### 2.3. The MBED Formulation

The objective of the MBED problem is to maximize the social welfare of the neighborhood of prosumers at every time step of the simulation, while satisfying the demand and allocating the power generation. The problem formulation as applied in this work for a time step $t$ takes the following form:

$$\min_{D} \quad \sum_{n \in \Omega} \left( C_n(P_n^t) + \hat{C}_n(\boldsymbol{p}_n^t) \right) \quad (7)$$

$$\text{s.t.} \quad \underline{P_n} \leq P_n^t \leq \overline{P_n} \qquad \forall n \in \Omega \quad (8)$$

$$P_{nm}^t + P_{mn}^t = 0 \qquad \forall (n,m) \in (\Omega, \omega_n) \quad (9)$$

$$P_{nm}^t \geq 0 \qquad \forall (n,m) \in (\Omega_p, \omega_n) \quad (10)$$

$$P_{mn}^t \leq 0 \qquad \forall (n,m) \in (\Omega_c, \omega_n) \quad (11)$$

with $D = (\boldsymbol{p}_n^t \in \mathbb{R}^{|\omega_n|})_{n \in \Omega}$; and $\Omega_p$ and $\Omega_c$ denoting the sets of producers and consumers, respectively. A prosumer is subject to inject a power quantity within its capabilities as stated in Equation (8). The role of consumer and producer is enforced by the sign of the power injected and that is reflected by constraints (10) and (11), respectively.

Since the MBED is a convex optimization problem, there is a single optimal solution. From here on, the followed methodology is described for solution through the RCI approach as in [7]. The approach followed in this work does not extend for the multi-time step formulation, in which ramping constraints and time-dependent behavior such as BESS charge/discharge can be considered. Here, BESS charge and discharge are performed within the prosumers' premises prior to the application of the P2P market mechanism. For such exemplary multi-time step formulation and solution of the C + I approach, the reader is referred to [15].

The stated problem is separated into subproblems subject to be solved locally by each prosumer. The cost function's parameters $a_n$ and $b_n$ as well as the bilateral trading coefficient $c_{nm}^g$ are private to every prosumer. Indeed, every market participant is bound by the trading reciprocity constraint only, formulated in Equation (9), as to what information is shared among prosumers in the market. The proposed convex optimization problem verifies the sufficient condition for strong duality called Slater's condition. The implications are that it is possible to solve this problem in a decentralized manner, where all prosumers focus only on solving their own local welfare maximization problem and it is only by construction that, by doing so, the maximum social welfare is achieved at convergence.

*2.4. The Relaxed Consensus and Innovations Algorithm*

At every time step, $t$, of the formulation at which a market is formed, each prosumer has to solve the following problem at iteration $i$:

$$\min_{D} \quad C_n(P_n^t + \hat{C}_n(p_n^t)) - (p_n^t)^\top \lambda_n^{i,t} \tag{12}$$

$$\text{s.t.} \quad \underline{P_n} \leq P_n^t \leq \overline{P_n} \tag{13}$$

$$P_{nm}^t \geq 0 \qquad \forall m \in \omega_m \ if \ n \in \Omega_p \tag{14}$$

$$P_{mn}^t \leq 0 \qquad \forall m \in \omega_m \ if \ n \in \Omega_c \tag{15}$$

where the Lagrangian dual variable $\lambda_n^{i,t} = (\lambda_{nm}^{i,t})_{m \in \omega_n}$ denotes the vector of price estimates of prosumer $n$ at iteration $i$ in time step $t$ and $(p_n^t)^\top$ is the transposed vector. Here, in order to solve this problem, a gradient step approach is taken and power boundary constraints are enforced through Lagrangian relaxation as in [16]. The dual variables for the complementary slackness of the power boundaries are denoted as $\underline{\mu_n}$ and $\overline{\mu_n}$. An iteration of the RCI algorithm as solved by prosumer $n$ is shown in Figure 2.

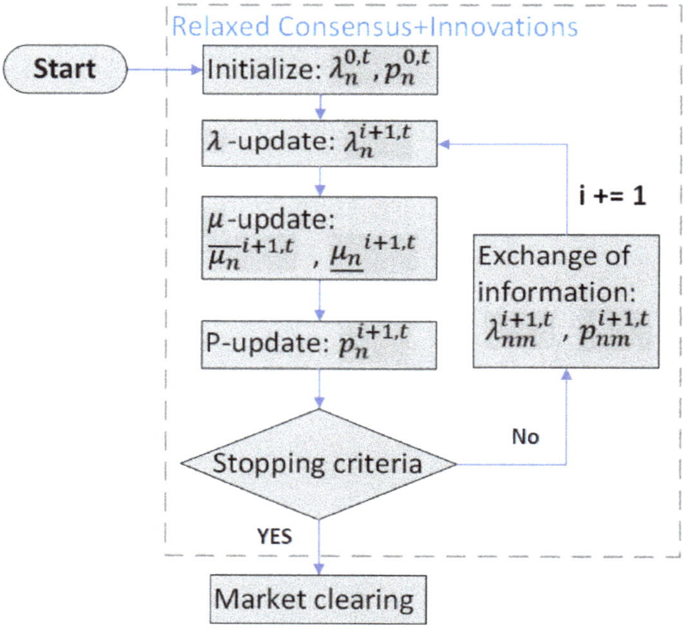


**Figure 2.** Diagram of the Relaxed Consensus + Innovations algorithm.

The RCI algorithm can be split into three main steps. The first and second steps are the updates for the dual variables $\lambda_n^{i,t}$ of the trading reciprocity constraint and the power boundary constraints $\underline{\mu_n}^{i,t}$ and $\overline{\mu_n}^{i,t}$. The third step is the update of the decision variables by using the gradient step method. The update of the price estimates is carried out by each prosumer, who comes up with a solution for each bilateral trade with the rest of prosumers. Although the calculation is performed locally, a consensus has to be accomplished on the price estimates for each pair of prosumers involved in a trade (i.e., $\lambda_{nm} = \lambda_{mn}$) after convergence. Similarly, the power reciprocity constraint in Equation (9) is enforced during the dual variable update such that the power quantity observed by the prosumers involved in the trade is equal ($P_{nm} = P_{mn}$). This is given by the following $\lambda$-update:

$$\lambda_{nm}^{i+1,t} = \lambda_{nm}^{i,t} - \beta^{i,t}(\lambda_{nm}^{i,t} - \lambda_{mn}^{i,t}) - \alpha^{i,t}(P_{nm}^{i,t} - P_{mn}^{i,t}) \tag{16}$$

where $\alpha^i$ and $\beta^i$ are sequences of positive factors according to [17]. These parameters are subject to tuning and constitute a trade-off between convergence speed and resilience to change. The dual variable update of the power boundary constraints is given by the following:

$$\overline{\mu_n}^{i+1,t} = max(0, \overline{\mu_n}^{i,t} + \eta^{i,t}(P_n^{i,t} - \overline{P_n^t})) \tag{17}$$

$$\underline{\mu_n}^{i+1,t} = max(0, \underline{\mu_n}^{i,t} + \eta^{i,t}(\underline{P_n^t} - P_n^{i,t})) \tag{18}$$

where $\eta^i$ is a persistent sequence positive tuning parameter similar to $\alpha^i$ and $\beta^i$. The power set point for a negotiated trade between any two agents $n$ and $m$, if $\hat{\lambda}_{nm}^t = \lambda_{nm}^t - c_{nm}^t$, can be written as Equation (19) and the decision variable is updated as in Equation (20):

$$P_n^{(m),i+1,t} = \frac{\hat{\lambda}_{nm}^t - \overline{\mu_n} + \underline{\mu_n} - b_n}{a_n} \tag{19}$$

$$P_{nm}^{i+1,t} = max(0, P_{nm}^{i,t} + f_{nm}^{i,t}(P_n^{(m),i+1,t} - P_n^{i,t})) \tag{20}$$

where $f_{nm}^{i,t}$ is an asymptotically proportional parameter with the form as in Equation (21). This is the primal update in the case of a prosumer acting as a generator. The $max$ operator in (17), (18) and (20) is to be substituted for a $min$ operator in the case of a prosumer acting as a load:

$$f_{nm}^{i,t} = \frac{|P_{nm}^t| + \delta^{i,t}}{\sum_{l\in\omega_n}(|P_{nl}^t + \delta^{i,t}|)} \tag{21}$$

where $\delta^{i,t}$ is a positive persistent sequence for which its value is taken from [7]. The iterative process is finished when the stopping criterion is met. Three parameters are defined in order to check for convergence of the primal and dual updated. Namely, $\epsilon_P$, $\epsilon_\lambda$ and $\epsilon_\mu$, as follows.

$$|\lambda_{nm}^{i+1,t} - \lambda_{nm}^{i,t}| \leq \epsilon_\lambda \tag{22}$$

$$|P_{nm}^{i+1,t} - P_{nm}^{i,t}| \leq \epsilon_P \tag{23}$$

$$|\mu_n^{i+1,t} - \mu_n^{i,t}| \leq \epsilon_\mu \tag{24}$$

The criterion in Equation (24) is optional but can provide better monitoring of the convergence. Here, only the dual price estimates and primal updates are inspected further and their implementation is recommended by [7], where more insights on the iterative process are found and from which the values of parameters $\alpha$, $\beta$, $\delta$ and $\eta$ are taken. Further details on these parameters are found in the Appendix.

The amount of information exchange per iteration is very low and restricted to the power and price estimates. Cost function parameters and power injection boundaries $(a_n, b_n, \overline{P_n}, \underline{P_n})$ as well as criterion parameter $(c_n^g, \gamma_{nm}^g)$ are kept private for each prosumer involved in a multi-bilateral trade, rendering this P2P market structure very secure and robust against an agent gaming the market.

A central coordinator is simulated, who checks for possible congestions, identifies the location of hazardous operation, stores the results of the network power flow, the information about the overall simulation process outside prosumer premises and shares relevant information with the prosumers. It is also another agent of the micro-grid and a P2P market participant capable of buying and selling energy. It represents the slack bus at the transformer with the external grid at the higher voltage level, which can absorb the surplus local power generation or deliver the needed power at moments where the demand exceeds the generation within the micro-grid.

For the purpose of information exchange, a broker entity is created to manage the information exchange and it is responsible for receiving all sent messages of all agents. Those messages (i.e., $P_{nm}$ and $\lambda_{nm}$) are sent to the exchange from where the broker forwards them to the rest of the peers. The events are ordered as follows: (i) The broker sends the common estimates at iteration $i$ to all agents in the market, (ii) all agents run an iteration of the RCI algorithm with the new information and yield their local updates at iteration $i+1$, (iii) agents send the information back to the broker and (iv) the stopping criterion is evaluated in order to break the iteration or to continue it.

### 2.5. Safety Operation of the Grid

The central coordinator has the permissions to apply an external cost to the prosumers in the grid. In this work, criterion parameters $\{c_n^g, \gamma_{nm}^g\}$ under criterion $g \in G$ are pushed to the central coordinator. Congestion is here defined as over-voltage at buses with a per-unit voltage $V[p.u.] \geq 1.03$ and thermal overload at lines with loading percent $\geq 80\,\%$.

In this work, three criteria, $g_1$, $g_2$ and $g_3$ are chosen to be over-voltage, thermal overload and the distance from a bus with over-voltage to the rest of buses where agents are connected to, respectively. Distance is referred to as the length of the cable that connects one bus to the other. For the sake of simplicity, it is assumed that the grid topology is a strongly connected graph where a vertex is representative of a bus. This holds true for networks in which every market participant is able to trade with every other participant who is not directly and physically connected to the former. However, the power traded should be always able to reach the respective trading peer. Referring to buses is similar to referring to agents connected to them as for the set notation in the coming lines.

Penalty tariff under criterion $g_1$ is always applied in conjunction with that under criterion $g_3$. That is because the intention is to shift the provision of power from prosumer $n$ causing bus over-voltage to the rest of prosumers demanding its power. However, in order for the algorithm to find an alternative, the cost should be differentiable. Therefore the differentiating criterion is chosen to be the distance in km. The respective criterion values for criteria $g_1$ to $g_3$ are in this work chosen to be $c_n^{g_1} = c^{g_1} = 0.5\,c€/\text{kWh}$, $c_n^{g_2} = c^{g_2} = 0.5\,c€/\text{kWh}$ and $c_n^{g_3} = c^{g_3} = 0.5\,c€(\text{km} \cdot \text{kWh})$.

The central coordinator holds the information of what tariffs are to be applied to what prosumers as penalizations for a behavior that results in grid congestion. The penalties take the following form.

$$\rho^t = c^{g_1} \cdot \gamma^{g_1,t} + c^{g_2} \cdot \gamma^{g_2,t} + c^{g_3} \cdot \gamma^{g_3,t} \tag{25}$$

$$\gamma^{g_1,t} = (\gamma_{nm}^{g_1,t})_{n \in \Omega_{g_1}, m \in \Omega_{n,g_1}} \tag{26}$$

$$\gamma^{g_2,t} = (\gamma_{nm}^{g_2,t})_{n \in \Omega_{g_2}, m \in \Omega_{n,g_2}} = (\gamma_{mn}^{g_2,t})_{n \in \Omega_{g_2}, m \in \Omega_{n,g_2}} \tag{27}$$

$$\gamma^{g_3,t} = (\gamma_{nm}^{g_3,t})_{n \in \Omega_{g_3}, m \in \Omega_{n,g_3}} \tag{28}$$

Once the algorithm converges and the market is cleared, the results of the power trades are passed to a power flow calculation instance in order to acknowledge the new state of the buses and lines, given the new power trades. Should the penalties not be great enough to shift the market trades towards a safety network operation, the penalization is updated in order to increase the magnitude of the tariffs imposed over problematic behavior. At a time step $t$ for an iteration $k$ in the global iteration, penalty updates are performed iteratively as suggested in [18] with the following form:

$$\rho_{nm}^{k+1,t} = c_n^{g,t}[\gamma_{nm}^{g,t} + (k-1)\Delta^{up}] \tag{29}$$

where $\Delta^{up}$ is an incremental coefficient to increase the tariff $\rho$ over iterations $k$, only activated after the second global iteration. If the grid state after the market clearance is proved to not incur in congestion at any point, the optimal power flow has been found, the global iteration is exited and the simulation can continue to the next time step $t+1$. The overall process runs by (i) checking whether the network power flows at time step $t$ incurs in grid congestion, (ii) identifying prosumers causing congestion, (iii) applying penalization to the respective prosumers at iteration k, (iv) running the RCI algorithm and obtaining a market clearing with traded power quantities, (v) running a power flow to check the new state of the grid and (vi) updating the penalties in iteration $k+1$ if congestions are not fixed or updating exit processes if the network is safely used and continued with $t+1$. Figure 3 illustrates the described process.

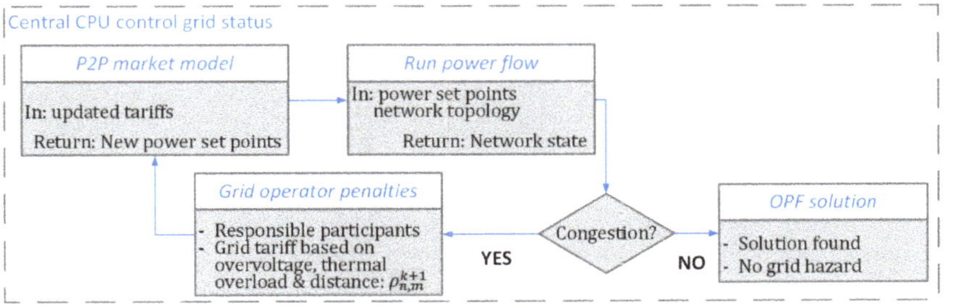


**Figure 3.** Diagram of central coordination control over secure network operation.

### 2.6. Limitations and Validation

Current methods for the solution of this optimization problem make use of the generator quadratic cost function model as the objective function and renewable energy sources such as PV are treated as must-take generation [17–21] and are accounted for as a system constraint or not accounted for at all [9,22,23]. Here, the quadratic generator cost function will be used as a virtual cost function to model prosumers' involvement in a P2P market and not necessarily to represent the cost of generation. This function serves the purpose of guiding the RCI algorithm towards an optimal solution that satisfies all demands while allocating the generated PV power production. The meaning of the linear and quadratic parameters $a_n$ and $b_n$ for prosumer $n$ is of key importance for allocating the production of that prosumer and whose magnitude directly influences the overall prices of electricity within the market. Should prosumers' have different cost functions, competition would drive variance on who sells or buys first from whom, as generation from cheaper installations is sold first (conversely, prosumers with lowest willingness to pay are first supplied).

The de facto implications of the prosumer's PV quadratic cost function parameters and the clearing prices that arise from them is subject to a variety of strategies for market configuration. Benefits of this approach are that it is possible to adjust parameters $a_n$ and $b_n$ for a number of prosumers in a micro-grid in order to obtain a clearing price matching the average wholesale price of electricity within a variation range. Moreover, setting these parameters as equal for all prosumers produces a scenario where prosumers with surplus PV generation are equally perceived by the RCI algorithm as for the incurred overall cost they introduce into the system.

Given the nature of renewable energy sources, the establishment of such P2P market is likely to fall into local power shortage or surplus due to PV power generation intermittency. In order to overcome this shortcoming, the external grid is modelled by the central coordinator as an additional P2P market participant representing the slack bus that acquires all prosumers' surplus power generation that could not be allocated locally in the micro-grid. Conversely, it delivers the needed power provided that the local prosumers' PV generation falls short to meet the local load demand. This behavior is represented by the following equation:

$$P_{ExtGrid}^{k,t} = -\sum_{n\in\Omega} P_n^{k,t} \quad (30)$$

Here, the generator system applies, being $P_n^{k,t} > 0$ for prosumers' PV power generation and $P_n^{k,t} \leq 0$ for prosumer demand. Therefore, the external grid agent buys surplus PV generation power when $P_{ExtGrid}^{k,t} \leq 0$ and sells power when $P_{ExtGrid}^{k,t} > 0$. This allows for the possibility to maintain a continuous P2P market mechanism throughout day and night. The external grid's cost function parameters are tuned to give preference to trading with renewable energies first and serve as a reserve that only ensures power balance. At the time of tariff imposition, the quantity absorbed by the external grid, $P_{ExtGrid}^{k,t}$, is also shortened through the external grid penalty incremental coefficient $\Delta_{ExtGrid}^{up}$ and follows the following form:

$$P_{ExtGrid}^{k+1,t} = [P_{ExtGrid}^{k,t} + (k-1)\Delta_{ExtGrid}^{up}] \tag{31}$$

$$\Delta_{ExtGrid}^{up} < 0 \qquad if P_{ExtGrid}^{k,t} > 0 \tag{32}$$

$$\Delta_{ExtGrid}^{up} > 0 \qquad if P_{ExtGrid}^{k,t} < 0 \tag{33}$$

$$\Delta_{ExtGrid}^{up} = 0 \qquad if P_{ExtGrid}^{k,t} = 0 \tag{34}$$

where $\Delta_{ExtGrid}^{up}$ takes negative values when the purpose is to reduce the external grid supply and positive values for when the aim is to reduce the external grid demand. The effect is that whenever there is a congestion occurrence and the corresponding tariff is applied, the external grid also reduces the absolute power injection in order to produce an actual shift in the demand.

## 3. Results

For the purpose of showing how the proposed methodology manages the interaction of different prosumers and solves congestion events during an optimal power-flow, a 12-Bus meshed network has been created. The topology of the network is shown in Figure 4. All prosumers count with PV installations and BESS. Moreover, they have the same cost function parameters which are smaller in magnitude than the cost function parameters chosen to represent external grid agents. The parameters for prosumers in the networks are provided in Table 1.

**Table 1.** Parameters for external grid and prosumer in selected 12-Bus Network.

| Bus | a | b | Installed_pv (kW) | Battery_Capacity (kWh) | Initial_SOC (%) |
|---|---|---|---|---|---|
| Ext grid | 0.06 | 35 | - | - | - |
| Bus LV0 | 0.05 | 25 | 9.9 | 3 | 30 |
| Bus LV1.1 | 0.05 | 25 | 5.1 | 2 | 30 |
| Bus LV1.2 | 0.04 | 25 | 5.1 | 2 | 30 |
| Bus LV1.3 | 0.05 | 25 | 3 | 1 | 30 |
| Bus LV1.4 | 0.05 | 25 | 13.8 | 4 | 30 |
| Bus LV1.5 | 0.05 | 25 | 13.8 | 4 | 30 |
| Bus LV2.1 | 0.05 | 25 | 2.1 | 1 | 30 |
| Bus LV2.2 | 0.05 | 25 | 3 | 1 | 30 |
| Bus LV2.3 | 0.05 | 25 | 4.8 | 2 | 30 |
| Bus LV2.4 | 0.05 | 25 | 8.7 | 3 | 30 |
| Bus LV2.2.1 | 0.05 | 25 | 8.1 | 3 | 30 |
| Bus LV2.2.2 | 0.05 | 25 | 9.9 | 3 | 30 |

The parameters chosen for the RCI are set to have dual variables convergence stopping criteria as provided in Table 2. The election of a high power convergence criterion comes with the trade-off of lower power allocation accuracy but a faster execution time and reduction in the average number of iterations needed to converge at every RCI process. The external grid penalty incremental $\Delta_{ExtGrid}^{up}$ is set to 10 kW. It is expected that the higher

$\Delta^{up}_{ExtGrid}$ is, the less surplus generation the grid accepts to buy and the higher the curtailment is in those moments where over-voltage at certain buses occur. Thermal overload is not found at any line as the power flow is not big enough at this low voltage level line for the given load demand and PV generation. For that reason, only over-voltage occurrence is used for the demonstration of the model capabilities.

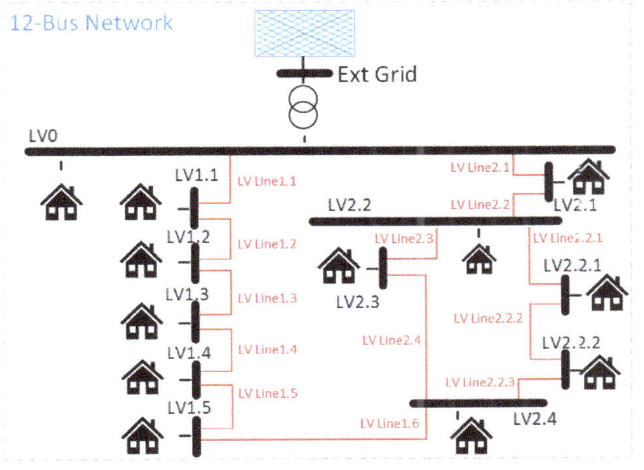


**Figure 4.** A 12-Bus LV Meshed distribution network consisting of prosumers.

**Table 2.** Dual variable RCI convergence stopping criteria. A 12-Bus Network.

| RCI Model Parameter | Value |
|---|---|
| $\epsilon_\lambda$ | 0.001 |
| $\epsilon_P$ | 0.01 |

The simulation is performed in a minute interval throughout the day and it is required to choose a demonstrative time step to show the effects of the applied methodology. Therefore, the time step 09:31 a.m. is chosen as the representative time step, where an over-voltage occurs in Bus LV2.2.2 corresponding to prosumer 12, which is shown in Figure 5. The vertical axis shows how much the per-unit voltage at each bus surpasses 1.0 p.u. From the figure, it is clear that before the application of RCI, the voltage at bus where prosumer 12 is connected is 1.033 p.u., which is reduced to 1.009 p.u. after the application of the RCI algorithm.

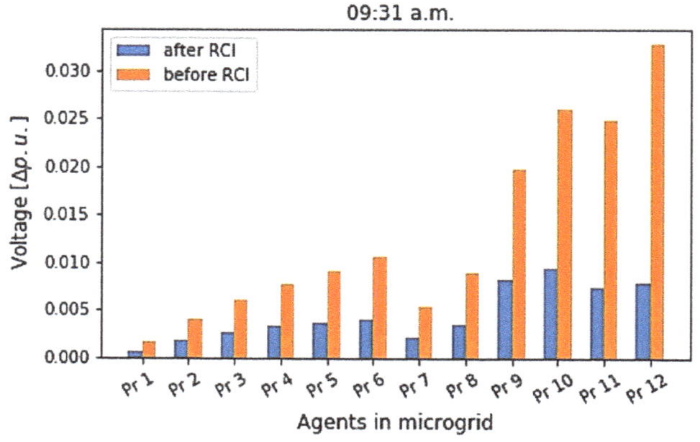


**Figure 5.** Voltage difference surpassing 1.0 p.u. at each prosumer bus at time step 09:31 a.m.

Figure 6 shows the power quantities traded by the prosumers and external grid agent at time step 09:31 a.m. Prosumer 12 sees its trades with the rest of the agents penalized with higher cost, shifting the external grid's demand preferably towards the rest of the prosumers first, as the demand is progressively satisfied. The demand of power from the external grid is shortened after the application of RCI algorithm by the application of the penalty tariff. This implies that external grid's demand is fully satisfied with less power purchase, leaving prosumer 12 completely out of the trading. The market condition permitted prosumer 12 to curtail full power quantity, which was available to sell.

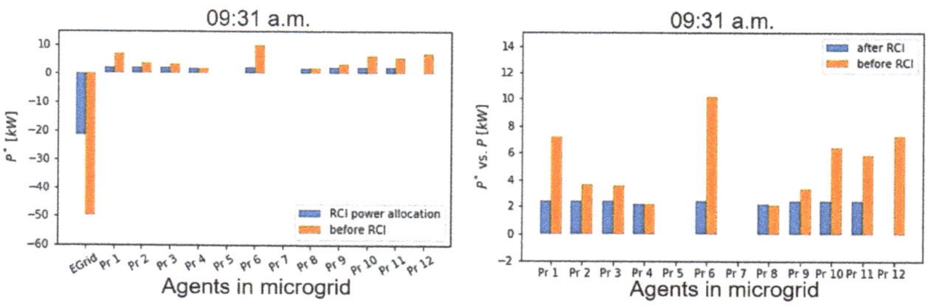


**Figure 6.** Power trades before and after RCI algorithm congestion correction. (**Left**): All agents are shown. (**Right**): Only prosumers' power set points. Time step: 09:31 a.m.

Since all prosumers account for the same cost function parameters, they are all observed by the RCI algorithm as equally eligible peers to trade with. That is why all prosumers have progressively sold the same amount of power to the external grid agent having a flat picture as drawn by the blue bars in Figure 6. As the RCI runs for the first time and a congestion is found, it is re-run with the mentioned penalties and external grid demands an update. The $\lambda$-convergence and the power convergence for non-zero trades are shown in Figure 7 from the perspective of external grid.

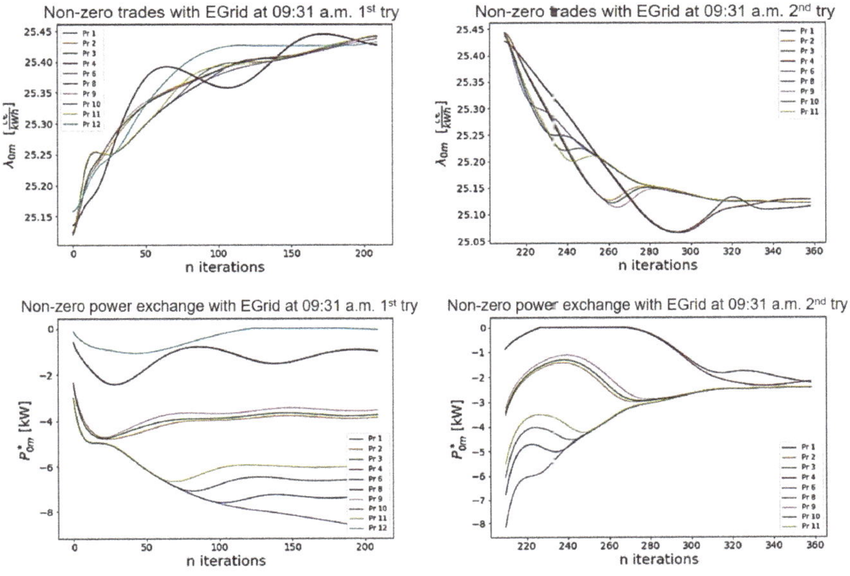


**Figure 7.** RCI $\lambda$-convergence and power convergence of an over-voltage occurrence at prosumer 12's bus. (**Left**): Over-voltage congestion occurs. (**Right**): Congestion is fixed. Time step: 09:31 a.m.

Each prosumer tried selling its full surplus power produced from PV and RCI algorithm yielded a power cost of approximately 25.25 $c€/kWh$. However, after the recognition of over-voltage occurrence and the respective penalty updates, the implemented RCI algorithm generates a new solution respecting the security of the network, which is shown at the right-hand side of Figure 7. An overview of the evolution of the voltage levels at the prosumer buses during the simulation horizon of 24 h is shown in Figure 8. The left hand side shows the voltages before the application RCI algorithm, where we can observe that the voltage levels increase beyond the allowed limits at mid-day, which is a time with high PV production. However such violations are removed after the application of the algorithm, as observed on the right-hand side.

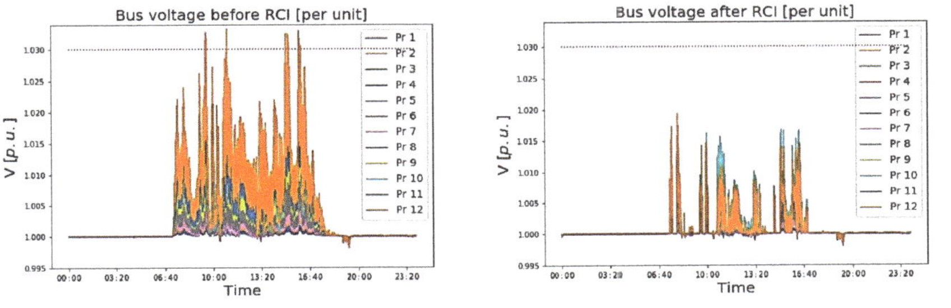


**Figure 8.** Voltage level evolution in per-unit throughout the entire simulation of one day for all buses. (**Left**): Over-voltage congestion occurs. (**Right**): Congestions are fixed.

Table 3 shows the performance of the algorithm when run throughout an entire day. Since the data are sampled in minute intervals, the n° of RCI problems solved is likely to be 1440. In other words, one RCI solution each minute in order to find a market clearing. However, as some minutes where found to incur in network congestion at some point, penalties are applied and the RCI needs to run again finding a new optimal market clearing. Hence, more iterations than amount of sample points in the data are needed.

**Table 3.** Results of a simulation of a day. A 12-Bus Network.

| Description | Value | Unit |
|---|---|---|
| Run Time | 7.6 | min |
| n° of RCI problems solved | 1452 | - |
| n° of RCI with >1 iteration, i | 581 | - |
| Average n° of RCI iterations for RCI with >1 iteration | 250.3 | - |
| Median of RCI iteration for RCI with >1 iteration | 40 | - |
| Average time per RCI solution | 0.31 | sec |

The load profiles vary little during some periods of the night and day. Therefore, the RCI algorithm requires no extra effort to find a solution from minute to minute when the state of the network is fairly stable, needing only 1 iteration to find the optimal allocation. Conversely, times with load or PV production variation need more RCI iterations to converge. Table 3 shows how many minutes of the day needed more than 1 RCI iteration, meaning higher power requirement variation. Average and median of the n° of RCI iterations show some minutes required a higher effort in order to find the optimal solution.

Figure 9 shows the a histogram of the number of iterations needed for each RCI Algorithm to converge, given that the RCI algorithm needed more than one iteration to find the solution. Linear scale on the left-hand side versus logarithmic scale on the right-hand side. Specifications on the equipment used in this work are found in the Table A1 in the Appendix B.

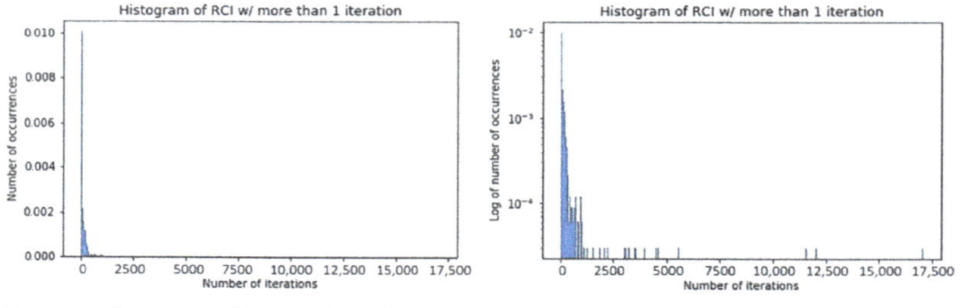


**Figure 9.** Histogram of the number of iterations for the convergence of each RCI process that needed more than one iteration to find a solution. (**Left**): Linear scale. (**Right**): Logarithmic scale.

## 4. Conclusions

A power flow calculation model has been developed in order to simulate a prosumer's behavior. The prosumer is expected to first satisfy its demand via its own PV installation to then charge the BESS with its eventual surplus power production and, finally, to feed the BESS rejected power into the local grid in exchange for monetary payment. Similarly, had the prosumer been unable to fully satisfy its demand throughout the PV installation and the BESS discharge, the prosumer is supplied via the local grid, in that order.

A P2P market coordination is established where prosumers are considered agents in the market along with the micro-grid operator, modelled as the EGrid agent, that also participates in the trade mechanism. The P2P market coordination is constructed via the RCI algorithm that solves the MBED problem at every time step of the simulation. The problem is solved in a decentralized manner and locally at each agent's premises. All agents focus only on solving their own local welfare maximization problem and it is only by construction that, by doing so, the maximum social welfare is achieved at convergence. Information exchange is finally limited to the exchange of the power quantities willing to be traded plus the price at which they ought to be sold or bought. This allows for a high-quality data protection trading scheme. A grid operator agent in the market is placed in charge of supervising the optimal and non-hazardous use of the network, continuously checking for congestion occurrence defined as over-voltages at certain buses and/or thermal overload at certain power transmission lines. This agent is allowed to impose a penalty tariff on each trade found to be incurring in congestion. Hence, the market is re-established with updated trading prices and the offers and demands are shifted to find the new optimal trades that simultaneously prevent congestion again.

### 4.1. Findings

The RCI model parameters allow for enormous possibilities as it accounts for several key parameters that incur in trade-offs between performance speed and convergence error. There is a wealth of trading schemes that can be constructed by tuning the parameters for the cost of power production and purchase that allows for great variation of the clearing prices, allowing for the simulation of different markets. Parameters related to the cost of generation/consumption per agent produce an immense amount of markets by selecting which agents ought to sell/buy first from which agents.

This work focuses in a renewable-only market, where penalties and rewards for participants can be based on a wealth of criteria such as tariff on congestion, energy origin preference, distance to source of power, etc. Furthermore, the criteria onto which penalties or rewards are settled can be pushed centrally or completely decentralize in order to allow agents to decide what they consider more costly or what reward's they are more prone to, enabling a complete product differentiation scheme in the latter scenario.

The communication graph that drives the P2P trading scheme is chosen here to be a strongly connected graph, i.e., every vertex of the graph is reachable from every other vertex of the same graph. Vertex is here referred to agents' points of connection. The graph can differ from the physical connection of agents through direct cable.

### 4.2. Future Work

Deeper investigation should be performed with dedicated computational characteristics and a real-world distributed system in order to better assess the algorithm speed. Moreover, the proposed algorithm can be itself implemented more optimally in terms of run-time by re-structuring the imperative flow. Further improvement should be conducted by testing asynchronous programming or parallelization in this regard. Real-world implementation should come with the need to asses possible lengthy response times after optimization from the part of the algorithm as well as the study of smaller simulation time steps.

Further work should be conducted where the proprietary BESS plays an active role in the OPF problem at a grid-level, optimizing the charge and discharge and making use of inter-temporal constraints and multi-time step market simulation. Insights on the possible application of such multi-step optimization are found, for instance, in [15].

An important tool that could be integrated with our proposed methodology to mitigate the grid stability issues with prosumers is the forecast of their stochastic power generation, especially at the distribution grid level. In this regard, several works such as [24,25] provide forecasting methodology for PV generation profiles, which could provide useful integration into the developed methodology here presented. This would in turn prove helpful in more accurate power flows, price convergence and improved grid stability.

Further investigations should be made on the asynchrony of the communication scheme as well as the intermittency of the connections among agents as performed in [26]. Moreover, the communication graph is supposed to be strongly connected. Further studies should be conducted on how to easily and quickly define changing environments as lines undergo failure, maintenance or any other event that affects the graph's characteristics.

Model parameters such as positive tuning parameters $\alpha^i$, $\beta^i$, $\delta^i$ and $\eta^i$ for the RCI should themselves be subject to further optimization and ought to be studied in more detail. This numerical optimization lies beyond the scope of this work. Furthermore, the magnitude of the cost criterion for the bilateral trading coefficient, $c^g_{nm}$, can produce different market outcomes. The election of these magnitudes can be further explored as to what criterion is of most interest given different market scenarios. The optimization problem of finding the right amount of penalization required in order to prevent a congestion occurrence is, in and of itself, a problem subject to deeper investigations.

**Author Contributions:** Conceptualization, D.K.W. and M.S.; methodology, D.K.W.; software, D.K.W. validation, D.K.W.; formal analysis, D.K.W.; investigation, D.K.W. and R.K.; resources, D.K.W.; data curation, D.K.W.; writing–original draft preparation, D.K.W., R.K.; writing–review and editing D.K.W., M.S., R.K.; visualization, D.K.W.; R.K., M.S.; project administration, M.S.; funding acquisition M.S. All authors have read and agreed to the published version of the manuscript.

**Funding:** This research was funded by Bundesministerium für Wirtschaft und Energie (BMWi) (FKZ 03EI6019E GaIN).

**Institutional Review Board Statement:** Not applicable.

**Informed Consent Statement:** No applicable.

**Data Availability Statement:** The data used for this work is discussed further down in the Appendix The source data is https://data.open-power-system-data.org (accessed on 24 April 2020).

**Conflicts of Interest:** The authors declare no conflict of interest.

## Abbreviations

The following abbreviations are used in this manuscript:

| | |
|---|---|
| ANM | Active Network Management |
| BESS | Battery Energy Storage System |
| DER | Distributed Energy Resources |
| EGrid | External Grid Agent |
| FPP | Federated Power Plant |
| MBED | Multi-Bilateral Economic Dispatch |
| OPF | Optimal Power Flow |
| P2P | Peer-to-Peer |
| PV | Photovoltaic |
| RCI | Relaxed Consensus + Innovations |
| SOC | State of Charge |
| VPP | Virtual Power Plant |

## Appendix A

The values of the tuning parameters of The RCI algorithm discussed in Section 2 are as follows.

$$\alpha^i = \frac{0.01}{\iota^{0.01}} \qquad \beta^i = \frac{0.1}{\iota^{0.1}} \qquad \delta^i = 1 \qquad \eta^i = 0.005 \tag{A1}$$

## Appendix B

The same data profile is used for all prosumers in order to represent that all belong to the same neighborhood where solar irradiation variance is negligible. PV generator data are given in kWh/min. Data are normalized to per-unit hour and are scaled up by multiplying the series times the prosumer's installed PV capacity. Figure A1 shows the PV profile as well as the load profiles used for prosumers, which are also taken from the same source.

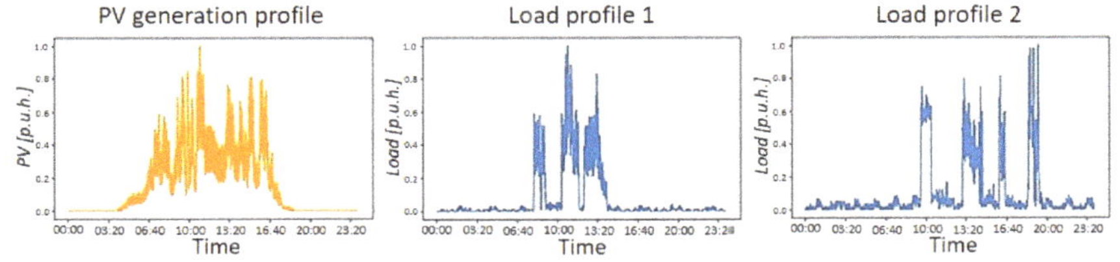


**Figure A1.** (**Left**): PV profile; (**Center**): Load profile 1; (**Right**): Load profile 2.

## Appendix C

The following Table A1 shows the specifications of the equipment that run the simulation here presented.

**Table A1.** Equipment used for the performance of the simulations.

| Component | Value |
|---|---|
| CPU | Ryzen 5 3600 4.2 GHz 35 MB cache |
| Memory | Crucial Ballistix Sport LT 16 GB (2 × 8 GB) 3000 MHz DDR4 |
| Power supply | Corsair RM850 |
| Motherboard | ASRock Socket AM4 m-ATX B450M PRO4 |

## References

1. Parag, Y.; Sovacool, B.K. Electricity market design for the prosumer era. *Nat. Energy* **2016**, *1*. doi:10.1038/nenergy.2016.32.
2. World Energy Council. *World Energy Trilemma Index 2016*; World Energy Council: London, UK, 2016.
3. Brusco, G.; Burgio, A.; Menniti, D.; Pinnarelli, A.; Sorrentino, N. Energy Management System for an Energy District With Demand Response Availability. *IEEE Trans. Smart Grid* **2014**, *5*, 2385–2393. doi:10.1109/TSG.2014.2318894.
4. Morstyn, T.; Farrell, N.; Darby, S.J.; McCulloch, M.D. Using peer-to-peer energy-trading platforms to incentivize prosumers to form federated power plants. *Nat. Energy* **2018**, *3*, 94–101. doi:10.1038/s41560-017-0075-y.
5. Cai, D.W.; Adlakha, S.; Low, S.H.; de Martini, P.; Mani Chandy, K. Impact of residential PV adoption on Retail Electricity Rates. *Energy Policy* **2013**, *62*, 830–843. doi:10.1016/j.enpol.2013.07.009.
6. Bronski, P. The Economics of Grid Defetion. *Rocky Mountain Institute*. 2014. Available online: https://rmi.org/ (accessed on 6 June 2020).
7. Sorin, E.; Bobo, L.; Pinson, P. Consensus-based approach to peer-to-peer electricity markets with product differentiation. Available online: https://arxiv.org/ (accessed on 31 May 2020).
8. Gill, S.; Kockar, I.; Ault, G.W. Dynamic Optimal Power Flow for Active Distribution Networks. *IEEE Trans. Power Syst.* **2014**, *29*, 121–131. doi:10.1109/TPWRS.2013.2279263.
9. Amini, M.H.; Jaddivada, R.; Mishra, S.; Karabasoglu, O. Distributed security constrained economic dispatch. In Proceedings of the 2015 IEEE Innovative Smart Grid Technologies—Asia (ISGT ASIA), Bangkok, Thailand, 3–6 November 2015; pp. 1–6. doi:10.1109/ISGT-Asia.2015.7387167.
10. Kargarian, A.; Fu, Y. System of Systems Based Security-Constrained Unit Commitment Incorporating Active Distribution Grids. *IEEE Trans. Power Syst.* **2014**, *29*, 2489–2498. doi:10.1109/TPWRS.2014.2307863.
11. Kargarian, A.; Fu, Y.; Li, Z. Distributed Security-Constrained Unit Commitment for Large-Scale Power Systems. *IEEE Trans. Power Syst.* **2015**, *30*, 1925–1936. doi:10.1109/TPWRS.2014.2360063.
12. Chakrabarti, S.; Kraning, M.; Chu, E.; Baldick, R.; Boyd, S. Security Constrained Optimal Power Flow via proximal message passing. In Proceedings of the 2014 Clemson University Power Systems Conference, Clemson, SC, USA, 11–14 March 2014; pp. 1–8. doi:10.1109/PSC.2014.6808131.
13. Mohammadi, J.; Hug, G.; Kar, S. Agent-Based Distributed Security Constrained Optimal Power Flow. *IEEE Trans. Smart Grid* **2018**, *9*, 1118–1130. doi:10.1109/TSG.2016.2577684.
14. Thurner, L.; Scheidler, A.; Schäfer, F.; Menke, J.H.; Dollichon, J.; Meier, F.; Meinecke, S.; Braun, M. pandapower—An Open Source Python Tool for Convenient Modeling, Analysis and Optimization of Electric Power Systems. *IEEE Trans. Power Syst.* **2018**, *33*, 6510–6521. doi:10.1109/TPWRS.2018.2829021.
15. Hug, G.; Kar, S.; Wu, C. Consensus + Innovations Approach for Distributed Multiagent Coordination in a Microgrid. *IEEE Trans. Smart Grid* **2015**, *6*, 1893–1903. doi:10.1109/TSG.2015.2409053.

16. Boyd, S. Distributed Optimization and Statistical Learning via the Alternating Direction Method of Multipliers. *Found. Trends Mach. Learn.* **2010**, *3*, 1–122. doi:10.1561/2200000016.
17. Kar, S.; Hug, G.; Mohammadi, J.; Moura, J.M.F. Distributed State Estimation and Energy Management in Smart Grids: A Consensus$+$Innovations Approach. *IEEE J. Sel. Top. Signal Process.* **2014**, *8*, 1022–1038. doi:10.1109/JSTSP.2014.2364545.
18. Orlandini, T.; Soares, T.; Sousa, T.; Pinson, P. Coordinating Consumer-Centric Market and Grid Operation on Distribution Grid. In Proceedings of the 2019 16th International Conference on the European Energy Market (EEM), Ljubljana, Slovenia, 18–20 September 2019; pp. 1–6. doi:10.1109/EEM.2019.8916247.
19. Xu, Y.; Zhang, W.; Hug, G.; Kar, S.; Li, Z. Cooperative Control of Distributed Energy Storage Systems in a Microgrid. *IEEE Trans. Smart Grid* **2015**, *6*, 238–248. doi:10.1109/TSG.2014.2354033.
20. Kar, S.; Moura, J.M.F.; Ramanan, K. Distributed Parameter Estimation in Sensor Networks: Nonlinear Observation Models and Imperfect Communication. *IEEE Trans. Inf. Theory* **2012**, *58*, 3575–3605. doi:10.1109/TIT.2012.2191450.
21. Woo, C.K.; Sreedharan, P.; Hargreaves, J.; Kahrl, F.; Wang, J.; Horowitz, I. A review of electricity product differentiation. *Appl. Energy* **2014**, *114*, 262–272. doi:10.1016/j.apenergy.2013.09.070.
22. Yang, S.; Tan, S.; Xu, J.X. Consensus Based Approach for Economic Dispatch Problem in a Smart Grid. *IEEE Trans. Power Syst.* **2013**, *28*, 4416–4426. doi:10.1109/TPWRS.2013.2271640.
23. Binetti, G.; Davoudi, A.; Lewis, F.L.; Naso, D.; Turchiano, B. Distributed Consensus-Based Economic Dispatch With Transmission Losses. *IEEE Trans. Power Syst.* **2014**, *29*, 1711–1720. doi:10.1109/TPWRS.2014.2299436.
24. Dellino, G.; Laudadio, T.; Mari, R.; Mastronardi, N.; Meloni, C.; Vergura, S. Energy production forecasting in a PV plant using transfer function models. In Proceedings of the 2015 IEEE 15th International Conference on Environment and Electrical Engineering (EEEIC), Rome, Italy, 10–13 June 2015; pp. 1379–1383. doi:10.1109/EEEIC.2015.7165371.
25. Bruno, S.; Dellino, G.; La Scala, M.; Meloni, C. A Microforecasting Module for Energy Management in Residential and Tertiary Buildings †. *Energies* **2019**, *12*. doi:10.3390/en12061006.
26. Mohammadi, J.; Hug, G.; Kar, S. Asynchronous distributed approach for DC Optimal Power Flow. In Proceedings of the 2015 IEEE Eindhoven PowerTech, Eindhoven, The Netherlands, 29 June–2 July 2015; pp. 1–6. doi:10.1109/PTC.2015.7232606.

energies

MDPI

Article

# Robust Multi-Step Predictor for Electricity Markets with Real-Time Pricing

Sachin Kahawala [1], Daswin De Silva [1,*], Seppo Sierla [2], Damminda Alahakoon [1], Rashmika Nawaratne [1], Evgeny Osipov [3], Andrew Jennings [1] and Valeriy Vyatkin [2,3]

1 Centre for Data Analytics and Cognition, La Trobe University, Bundoora, VIC 3083, Australia; sachin.kahawala@ltu.se (S.K.); D.Alahakoon@latrobe.edu.au (D.A.); B Nawaratne@latrobe.edu.au (R.N.); Andrew.Jennings@latrobe.edu.au (A.J.)
2 Department of Electrical Engineering and Automation, School of Electrical Engineering, Aalto University, FI-00076 Espoo, Finland; seppo.sierla@aalto.fi (S.S.); valeriy.vyatkin@aalto.fi (V.V.)
3 Department of Computer Science, Electrical and Space Engineering, Luleå Tekniska Universitet, SE-97187 Luleå, Sweden; Evgeny.Osipov@ltu.se
* Correspondence: d.desilva@latrobe.edu.au; Tel.: +61-3-9479-6468

**Abstract:** Real-time electricity pricing mechanisms are emerging as a key component of the smart grid. However, prior work has not fully addressed the challenges of multi-step prediction (Predicting multiple time steps into the future) that is accurate, robust and real-time. This paper proposes a novel Artificial Intelligence-based approach, Robust Intelligent Price Prediction in Real-time (RIPPR), that overcomes these challenges. RIPPR utilizes Variational Mode Decomposition (VMD) to transform the spot price data stream into sub-series that are optimized for robustness using the particle swarm optimization (PSO) algorithm. These sub-series are inputted to a Random Vector Functional Link neural network algorithm for real-time multi-step prediction. A mirror extension removal of VMD, including continuous and discrete spaces in the PSO, is a further novel contribution that improves the effectiveness of RIPPR. The superiority of the proposed RIPPR is demonstrated using three empirical studies of multi-step price prediction of the Australian electricity market.

**Keywords:** demand response; real-time pricing; prosumers; electricity price forecasting; particle swarm optimization



**Citation:** Kahawala, S.; De Silva, D.; Sierla, S.; Alahakoon, D.; Nawaratne, R.; Osipov, E.; Jennings, A.; Vyatkin, V. Robust Multi-Step Predictor for Electricity Markets with Real-Time Pricing. *Energies* **2021**, *14*, 4378. https://doi.org/10.3390/en14144378

Academic Editor: Ricardo J. Bessa

Received: 14 June 2021
Accepted: 14 July 2021
Published: 20 July 2021

**Publisher's Note:** MDPI stays neutral with regard to jurisdictional claims in published maps and institutional affiliations.



## 1. Introduction

The global transition to renewable power generation has resulted in significant research efforts to design real-time approaches for power dispatch in power grids [1] and microgrids [2]. Real-time pricing is emerging as a solution for coordinating renewable generation with other intelligent energy resources [3], such as flexible loads [4], battery storages [5] and electric vehicles [6]. Several authors mean real-time pricing when they use the term 'demand response' [7]. In some works, real-time pricing refers to varying hourly prices that are determined day-ahead [8] or at the end of the day [9]. Anand and Ramasubbu [10] proposed an isolated microgrid with hourly changing real-time prices known only one hour in advance. However, a move towards real-time pricing with prices being determined one interval at a time at 5-min intervals offers powerful tools for retailers and utilities to coordinate the diverse, intelligent distributed energy resources of their customers [11]. The transformation of residential and commercial buildings into prosumers with local renewable generation is one driver for such short interval real-time pricing markets [12]. Elma et al. [13] proposed a domestic prosumer operating at five min intervals, rescheduling or curtailing loads according to forecasted local photovoltaic generation and real-time electricity prices. Mbungu et al. [14] presented a similar approach for a commercial building prosumer with photovoltaic generation and battery storage; the proposed real-time pricing scheme is built on top of a time-of-use pricing scheme. Mirakhorli and

Dong [15] demonstrated that a commercial building prosumer operating under five-minute real-time pricing could achieve major electricity cost savings in comparison to time-of-use or hourly pricing. Li et al. [3] optimize a multi-energy prosumer community in a market environment with real-time prices for electricity and district heating. In some regions electricity spot markets support real-time trading at 5 min intervals [16]. An example of such a market is the Australian spot market [17].

Alahyari and Pozo [18] presented an approach for maximizing the profits for electricity consumers participating in a demand response program. A real-time electricity price is assumed so that the price for the next hours is not known at the time of planning the demand response actions. The proposed framework is able to use a forecast of such a real-time price and is able to cope with uncertainties in the forecast. Thus, the approach presented in this paper could be directly exploited in the demand response optimization proposed by Alahyari and Pozo [18]. Moving to real-time spot prices, such as prices that change every 5 min, motivates a rethinking of energy management approaches to address a real-time timeframe. For example, weather forecasts that are crucial to consumption forecasting are usually not performed short-term to address weather disturbances. However, such a short-term forecast is provided in Thilker et al. [19]. These forecasts are advantageously used by a model-predictive controller managing indoor climate, with the goal of reducing electricity consumption while maintaining indoor comfort within specifications. Our real time electricity price forecast is not directly comparable to the short-term weather forecast in [19], as the price remains constant for the market interval, e.g., 5 min. However, as short intervals such as 5 min become more common in real-time electricity pricing, a rethinking of energy management system research to exploit short-term generation and consumption forecast will be needed.

This article proposes a novel real-time electricity price predictor, and the Australian spot market will be used as a case study due to the availability of open data. However our proposed approach uses generally applicable time series forecasting techniques that are not specific to spot markets, so the proposed forecasting method is adaptable to other real-time electricity markets such as those referenced above.

Time series forecasting is a mature field of study with diverse applications in academic industrial and business contexts. It is defined as the formulation of forecasts on the basis of data in one or more time series, where time series is a collection of observations made sequentially through time [20]. A forecasting method is distinguished from a forecasting model which takes into account underlying distributions of a time series. A forecast is predicated on the current time step, forecast horizon, and evaluated using the residual forecast error. In EPF, time series forecasting methods can be grouped into three categories statistical, machine learning and hybrid methods. Statistical methods are effective at capturing seasonality, machine learning captures non-linear behaviors of a time series such as sudden bursts or jumps, and hybrid methods break down the raw data stream into sub-components and then apply either statistical or machine learning methods on these components. Although hybrid methods exhibit high accuracy, they have only been demonstrated in theoretical settings, and this limits its value in addressing the practical challenges of balancing high accuracy with robust, real-time processing.

In this paper, we propose a new EPF method, Robust Intelligent Price Prediction in Real-time (RIPPR), to address these practical challenges. RIPPR is an ensemble technique that uses Variation Mode Decomposition (VMD) to decompose time series data streams into K sub-series, where K is chosen by particle swarm optimization (PSO) considering both forecasting accuracy and forecasting horizon. Each sub-series is modeled using a variant of Random Vector Functional Link (RVFL) neural networks, Extreme Learning Machine (ELM), for the h-step ahead point forecast. Finally, the h-step forecast for the given data stream is taken by aggregating the forecasted values for each sub-series.

The research contributions of this paper are as follows:

1. The design and development of RIPPR, a novel EPF ensemble using VMD and RVFL

2. Optimization of the VMD module using PSO to determine optimal modes of decomposition with respect to forecast accuracy and forecast horizon;
3. Extending the VMD module to process signal edges for real-time EPF applications;
4. Evaluation of RIPPR on three benchmark datasets and one real-world dataset, using metrics of accuracy and robustness. The four datasets are from diverse energy market settings that are representative of the complexities of EPF and the robustness of the proposed method.

The rest of the paper is organized as follows; Section 2 presents related work in statistical, machine learning and hybrid methods, followed by the proposed ensemble approach for EPF. The experiments and results are presented in Section 3, and Section 4 concludes the paper.

## 2. Materials and Methods

### 2.1. Related Work

Most related work in the domain of EPF is based on statistical models that derive underlying statistical properties of the time-series data streams for the task of forecasting. Some of the examples for statistical methods are autoregressive–moving average (ARMA), autoregressive integrated moving average (ARIMA), vector autoregression (VAR), Kalman filter-based methods, Holt–Winters exponential smoothing and generalized autoregressive conditional heteroskedasticity (GARCH). Chujai et al. [21] validated the capabilities of both ARMA and ARIMA in household electric consumption forecasting. Furthermore, they evaluate using the most suitable forecasting period for the given use case. Carolina et al. [22] used the VAR forecasting model to apply to interval time series. Girish et al. [23] presented the GARCH-based one-hour-ahead price forecasting model and empirically validated it using voluminous time series generated by the electricity market of India. The main limitation of statistical methods is the inability to detect or represent the non-linear features and random changes in a time series.

In contrast, EPF based on machine learning methods such as support vector machine (SVM), artificial neural networks (ANN), fuzzy neural networks (FNN), recurrent neural networks (RNN) and randomly connected neural networks is able to capture and represent these non-linear features. Ziming et al. [24] proposed a month ahead of daily electricity price profile forecasting based on SVM; SVM is adopted to forecast the prices of peak hours in peak months. Furthermore, they validated its effectiveness using the Electric Reliability Council of Texas (ERCOT). Anand et al. [25] deployed an ANN-based PSO model to forecast future energy demand for a state of India. Both particle swarm optimization (PSO) and Genetic algorithm (GA) were developed in linear and quadratic forms, and the hybrid ANN models were applied to different series. They have empirically evaluated the results comparing with other methods such as ARIMA, linear models. From the optimization perspective, they have validated the gains of the PSO-based model over the GA-based model. Yunpeng et al. [26] proposed a model for multi-step ahead time series forecasting using long short-term memory (LSTM) RNN. Hassan et al. [27] proposed a novel model based on randomly connected RNNs for electricity load demand forecasting, and the results prove the superiority of the proposed model. Compared to statistical methods, machine learning methods capture the non-linear features and random changes to a certain extent and maintains the potential for further improvements.

A separate stream of related work has focused on hybrid models composed of one or more statistical and machine learning techniques, as single models cannot effectively extract features from a complex time series such as those in energy markets that fluctuate rapidly. Hybrid models use different data decomposition techniques to process the non-linear and non-stationary electricity-related data before applying it to the forecasting model. Wang et al. [28] proposed a novel method that uses wavelet packet transform (WPT) to decompose the time series data and particle swarm optimization based on simulated annealing (PSOSA) and Least Square Support Vector Machine (LSSVM) for wind speed forecasting and the experiments demonstrated that the WPT decomposition technique makes great improvement on the forecast accuracy. Wang et al. [29] proposed a hybrid model that consists of a two-layer decomposition technique which includes fast ensemble empirical mode decomposition (FEEMD) and Variational mode decomposition (VMD). Further, the model uses back propagation (BP) neural network optimized by the firefly algorithm (FA) as the prediction algorithm. Yang et al. [30] proposed a multi-step electricity price forecasting algorithm based on the VMD algorithm, improved multi-objective sine cosine algorithm (IMOSCA), and regularized extreme learning machine (RELM). Additionally, they ensured the model is not dependent on new information during the testing phases, thereby increasing its practical value. Kaijian et al. [31] developed a method for forecasting electricity market risk using Empirical Mode decomposition (EMD) based on the Value at Risk (VaR) model, with Exponential Weighted Moving Average (EWMA) representing individual risk factors. Separately, decomposition-based TSF methods such as a multi-objective optimization for short-term wind speed forecasting [32], an ensemble empirical mode decomposition based crude oil price forecasting [33], as well as AI-based models that use deep recurrent neural networks [34], long short term memory networks [35], and hybrid neuro-fuzzy inference [36] for energy consumption prediction were reported in the recent literature.

Despite hybrid models reporting improvements to the accuracy and prediction horizon of time series forecasts, two major limitations are inherent in the development of such models. Firstly, the use of a fixed number of components for the decomposition of the raw time-series into train and test sets, which implies the test set is required in advance in the data pre-processing stage [30]. This means the model will underperform when deployed in a real-world setting where data is acquired in a sequential manner and cannot be decomposed in advance. Additionally, the model will not be able to adapt to any changes in the data stream. Secondly, decomposition has to be conducted at the arrival of each new data point. If the time step (time between two adjacent data points) is smaller than the time taken to decompose and forecast, such models become impractical for real-world application settings.

### 2.2. Proposed Method

The proposed method, RIPPR, is a machine learning ensemble-based decomposition method that addresses these limitations. In brief, the proposed approach consists of five main components. The pre-processing module includes a normalization as well as an extreme outlier removal process, which is then processed by the data decomposition module. The data decomposition module decomposes a given data stream into K sub-series where the optimal parameters for the decomposition are chosen by the optimization module, including the value K. It is followed up by the Forecasting module where each sub-series is modeled with RVLF for h-step ahead point forecast, which then aggregated for each subseries in the post-processing module to produce the h-step ahead point forecast. The RIPPR process is illustrated in Figure 1. It comprises of five modules, data pre-processing, data decomposition, optimization, time series forecasting and post-processing. Each module is delineated in the following subsections.

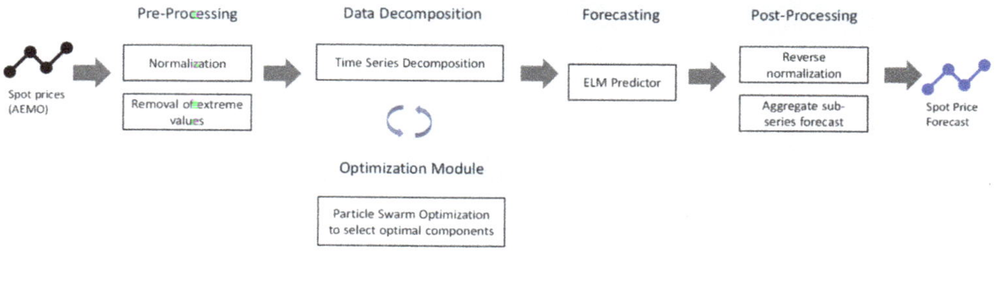


**Figure 1.** The Proposed Method-Robust Intelligent Price Prediction in Real-time (RIPPR).

### 2.2.1. Data Pre-Processing Module

The pre-processing module receives the raw time series data as input. In the context of energy markets, short-term EPF is a core capability of an energy market that drives the market's operational activities. The short-term EPF is also called spot or day-ahead price forecasting. Here we consider raw time series data to be the spot prices that the National Electricity Market Operators use to match the supply of electricity from power stations with real-time consumption by households and businesses. All electricity in the spot market is bought and sold at the spot price.

In general, to obtain an accurate forecast, the input time series data that are used to model the forecasting model should be normalized in consideration of the new data that the model will account for in the future. Due to the high fluctuation and varying nature of the energy market, each dataset and data sample is unique, posing unique challenges for EPF. In the context of spot prices, the primary challenge is the presence of noise, including duplicated values, missing data points, and extreme outliers that will make the forecasting model weak. In RIPPR, we adopt two techniques to suppress the noise in input data streams. First, we remove the extreme values to discard extreme outliers in the input data, and second, we normalize the input data prior to feeding it to the prediction model.

Extreme values (or outliers) are data points that significantly differ from other observations, and the removal of such extreme values is considered as one of the significant steps in data pre-processing. This is because machine learning algorithms and corresponding predictions/forecasts are sensitive to the range and distribution of the input data points; therefore, outliers can mislead the training process resulting in longer training times and less accurate models. Extreme values can be of two types, (1) outliers that are introduced due to human or mechanical errors, and (2) extreme values that are caused by natural variations of a given distribution. In the context of smart grid/spot prices, the first type is rarely attested. However, a common case is the presence of extreme outliers. For instance, wholesale energy prices are influenced by a range of factors, including weather, local economic activities, international oil prices and resource availability. The availability of such factors could lead spot prices to be extremely volatile and unpredictable. Thereby, we intend to address these extreme values using extreme value analysis that use the statistical tails of the underlying distribution of the variable and find the values at the extreme end of the tails. Followed by the extreme value removal, we perform min–max normalization on the time series data to scale the time series data in the range 0 and 1. In general, the min–max normalization technique does not handle outliers and extreme values, and this is why normalization is preceded by extreme value removal.

A limitation of the min-max normalization technique is that the values used in the train-test phases can be very different from a real-world scenario, where the minimum and maximum values of a time series is not prior. It is necessary to make a realistic assumption of the min–max values based on expert knowledge of the energy market.

### 2.2.2. Data Decomposition Module

Time series data can exhibit a variety of patterns; therefore, splitting such time series data into several distinct components, each representing an underlying pattern category, could lead to better analysis and pattern identification. The complex characteristics of the electricity spot price market make it even harder to capture the underlying patterns in order to forecast spot prices, which makes decomposition an essential component of the proposed approach. In recent work, a number of signal decomposition algorithms that can be utilized for time series forecasting were proposed. For example, Empirical Mode Decomposition (EMD) [37], Ensemble EMD [38], Complete Ensemble EMD with adaptive noise [39], Empirical Wavelet Transform (EWT) [40] and Variational Mode Decomposition [41] are several recent signal decomposition techniques.

As stated by Wang et al. [42], Variational Mode Decomposition (VMD) is the state-of-the-art data decomposition method in signal modeling. VMD decomposes a signal into an ensemble of band-limited Intrinsic Mode Functions (IMF). It is more effective than other signal decomposition methods as it is able to generate IMF components concurrently using the ADMM optimization method [43], it can avoid the error caused during the recursive calculating and ending effect, which is a significant issue of EMD [30] and it is significantly robust to noise as well [41].

In VMD, a real-valued input signal $s$ is decomposed into a discrete number of modes $u_k$ that have specific sparsity properties while reproducing the input. Each mode of $\chi_k$ is assumed to be most compact around a center pulsation $\omega_k$, which is determined along with the decomposition. Based on the original algorithm, the resulting constrained variational problem is expressed as follows.

$$min_{\{u_k\},\{\omega_k\}}\left\{\sum_k \left\| {}_t\left[\left(\delta(t)+\frac{j}{\pi t}\right) * u_k(t)\right]e^{-j\omega_k t}\right\|_2^2\right\} \text{s.t. } \sum_k u_k = f \quad (1)$$

where $\{u_k\}:=\{u_1,\ldots,u_k\}$ and $\{\omega_k\}:=\{\omega_1,\ldots,\omega_k\}$ are shorthand notations for the set of all modes and their center frequencies, respectively, and $f$ is the input signal. Equally $\sum_k := \sum_{k=1}^{K}$ is understood as the summation over all modes. Here, $K$ is the total number of the decomposed modes. Since the decomposition is mainly based on the parameter $K$, a significant effort should be placed to select the optimal value.

To address the constrained variational problem, VMD uses an optimization methodology called ADMM [41] to select the central frequencies and intrinsic mode functions centered on those frequencies concurrently. First, minimization with respect to $u_k$ (modes) is considered, and the following is obtained for $\hat{u}_k{}^{n+1}$:

$$\hat{u}_k^{n+1}(\omega) = \frac{\hat{f}(\omega) - \sum_{i\neq k}\hat{u}_i(\omega) + \frac{\hat{\lambda}(\omega)}{2}}{1+2\alpha(\omega-\omega_k)^2} \quad (2)$$

Secondly, minimization with respect to $\omega_k$ (center frequencies) is considered and following is obtained for $\omega_k{}^{n+1}$:

$$\omega_k^{n+1} = \frac{\int_0^\infty \omega|\hat{u}_k(\omega)|^2\, d\omega}{\int_0^\infty |\hat{u}_k(\omega)|^2\, d\omega} \quad (3)$$

Here $u_k{}^{n+1}$, $\omega_k{}^{n+1}$ and $\lambda^{n+1}$ are updated continuously until convergence. When the following convergence condition is met, the algorithm terminates, producing the K modes

$$\sum_k \frac{||\hat{u}_k^{n+1} - \hat{u}_k^n||_2^2}{||\hat{u}_k^n||_2^2} < \varepsilon \quad (4)$$

The generic VMD algorithm is effective for discrete, finite time signals; however, the boundaries of the signal are a key technical challenge due to the vanishing derivatives in the time domain boundary [41]. To address this challenge, VMD introduces a mirror extension of the signal by half its length on each side. However, this means the prediction is based on using previously seen values as future point forecasts. This is because decomposed sub-signals assume that the original signal will continue in the form of a mirror extension. Therefore, generic VMD cannot be used directly in a real-world time series forecasting setting. In RIPPR, we modified the VMD algorithm by removing this mirror extension.

In Figure 2, we compared the generic VMD algorithm and the modified version (that has the mirror extension removed) on a benchmark dataset. The results indicate that the two versions obviously differ, which will lead to different forecasting performances. However, the effectiveness of the modified-VMD algorithm is necessary for practical use

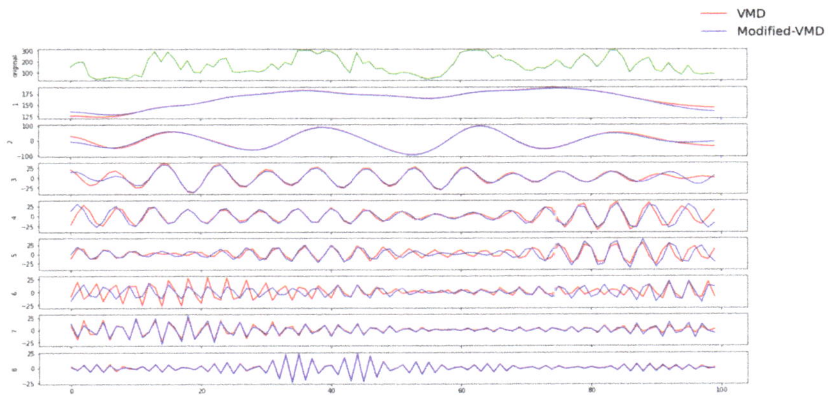


**Figure 2.** Data decomposition comparison between VMD and modified-VMD.

Returning to the core capability of the VMD method, the decomposition of a signal depends on the settings of its input parameters. The VMD method consists of five parameters, namely, mode number ($K$-the number of modes to be recovered), balancing parameter ($\alpha$-the bandwidth of extracted modes (low value of $\alpha$ yields higher bandwidth)), time-step of dual ascent ($\tau$), initial omega ($\omega$) and tolerance ($\varepsilon$). As experimentally proven by Dragomiretskiy and Zosso [41], $\varepsilon$, $\tau$ and $\omega$ has standard values across any given signal distribution. The standard values are; $\varepsilon = 1 \times 10^{-6}$, $\omega = 0$, $\tau = 0$. However, $k$ and $\alpha$ depends on the signal, and this means for each new signal distribution, these two parameters needed to be adjusted. We address this in the next module using particle swarm optimization (PSO).

2.2.3. Optimization Module

The number of modes to be recovered ($K$) and the balancing parameter ($\alpha$) determine the accuracy of the VMD decomposition. In this module, we utilize particle swarm optimization [44] (PSO) to select the most suitable values for these two values $K$, $\alpha$, for a given forecasting horizon. We consider the prediction time for a given time-step as the objective function of the optimization technique.

PSO is a metaheuristic parallel search technique used for the optimization of continuous non-linear problems, inspired by the social behavior of bird flocking and fish schooling [45]. PSO is a global optimization algorithm for addressing optimization problems on which a point or surface in an n-dimensional space represents the best solution. In this algorithm, several cooperative agents are used, and each agent exchanges information obtained in its respective search process. Each agent, referred to as a particle, follows two rules, (1) follow the best performing particle and (2) move toward the best conditions found by the particle itself. Thereby, each particle ultimately evolves to an optimal or a near-optimal solution. PSO requires only primitive mathematical operators and is computationally inexpensive in terms of both memory requirements and speed when compared with other existing evolutionary algorithms [46].

The standard PSO (Algorithm 1) algorithm can be defined using the following equations,

$$v_i(k+1) = \omega v_i(k) + c_1 r_1 .(p_{best,i} - x_i(k)) + c_2 r_2 .(g_{best} - x_i(k)) \tag{5}$$

$$x_i(k+1) = x_i(k) + v_i(k+1)\alpha \tag{6}$$

where $x_i$ is the position of particle $i$; $v_i$ is the velocity of particle $i$; $k$ denotes the iteration number; $\omega$ is the inertia weight; $r_1$ and $r_2$ are random variables uniformly distributed within (0, 1); and $c_1$, $c_2$ are the cognitive and social coefficient, respectively. The variable $p_{best,i}$ is used to store the best position that the $i$th particle has found so far, and $g_{best}$ is used to store the best position of all the particles. The basic PSO is influenced by a number of control parameters, namely the dimension of the problem, number of particles, step size ($\alpha$), inertia weight ($\omega$), neighborhood size, acceleration coefficients, number of iterations ($\text{iter}_{\text{max}}$), and the random values that scale the contribution of the cognitive and social components. Additionally, if velocity clamping or constriction is used, the maximum velocity and constriction coefficient also influence the performance of the PSO.

**Algorithm 1** Standard particle swarm optimization

**Input:** Objective function to be minimized (or maximized)
**Parameters:** swarm size, $c1$,$c2$,$\omega$, $\text{iter}_{\text{max}}$,error
**Output:** $\text{g}_{\text{best}}$
1: Initialize population (Number of particles = swarm size) with random position and velocity;
2: Evaluate the fitness value of each particle. Fitness evaluation is conducted by supplying the candidate solution to the objective function;
3: Update individual and global best fitness values ($\text{p}_{\text{best,i}}$ and $\text{g}_{\text{best}}$). Positions are updated by comparing the newly calculated fitness values against the previous ones and replacing the $\text{p}_{\text{best,i}}$ and $\text{g}_{\text{best}}$, as well as their corresponding positions, as necessary;
4: Update velocity and position of each particle in the swarm, using Equations (5) and (6);
5: Evaluate the convergence criterion. If the convergence criterion is met, terminate the process; if the iteration number equals $\text{iter}_{\text{max}}$, terminate the process; otherwise, the iteration number will increase by 1 and go to step 2.

A novel contribution of this module is that we have extended the basic PSO algorithm to take both continuous space ($\mathbb{R}^+$-space) and discrete space ($\mathbb{Z}^+$-space) for optimization. In the given context, two variables exist for the optimization purpose, namely $K$ and $\alpha$. The variable $\alpha$ is a continuous variable, while $K$ is a discrete variable. Therefore, we modify the basic PSO to consider both $\mathbb{R}^+$ and $\mathbb{Z}^+$ spaces in optimization.

At the start of the algorithm, we place particles randomly such that particle position for each particle with respect to $K$ is discrete. Then, we round off the $v_i(k+1)$ $\alpha$ to the nearest integer before adding it to $x_i$ $(k)$ (Equation (6)). As such, we change Equation (6) for variable $K$ as follows:

$$x_i(k+1) = x_i(k) + [v_i(k+1)\alpha] \tag{7}$$

where '[ ]' operation represents rounding to the nearest integer.

The following section describes the fitness function that is used in the RIPPR approach. This fitness function is selected to cover both prediction accuracy as well as time taken to the prediction. The more obvious fitness function will be to use the $\text{test}_{\text{RMSE}}$ directly so that PSO will find an optimal ($K$, $\alpha$) combination so that the forecasting accuracy will be higher. However, our experiments show that by doing so, it will result in a higher $K$ value which is not desirable when considering the time taken for the prediction ($K$ separate models will be created for each sub-series).

To overcome the aforementioned issue, we have included a penalty term to penalize having a higher $K$ value while having good accuracy. The final fitness function is as follows:

$$\text{Fitness function } = \min\{test_{rmse} + \beta \times K\} \tag{8}$$

where $\beta$ is constant, we can control the penalizing term by adjusting the $\beta$ value. From our experiments on energy price forecasting, we see that having $\beta$ = 1 leads to better accuracy as well as manages to penalize having a higher $K$ value precisely. Depending on the application, the value for $K$ should be chosen accordingly. The calculation of the fitness function is given in Algorithm 2.

**Algorithm 2** Fitness value calculation for PSO

**Input:** K, α, Data (X), forecasting horizon
**Output:** Fitness value
1: Decompose the data (X) using VMD for the given (K, $\alpha$) combination;
2: Divide each sequence (sub-series) into multiple input/output patterns called samples for the given forecasting horizon;
3: Split the samples set into train and test split at a ratio of 6:4;
4: Train on the train data using the time series forecasting module for each sub-series;
5: Predict for the test data using trained models for each sub-series;
6: Aggregate the predicted values for each sub-series to obtain the final prediction for the test data;
7: Calculate the RMSE value between actual values and predicted values for the test data ($test_{RMSE}$);
8: Calculate fitness value as follows: fitness value $= test_{rmse} + \beta \times K$.

In Figure 3, we illustrate the learning process of PSO to find the optimal components for VMD. This experiment is conducted using dataset A (Table 1). We used the following parameters in the PSO algorithm, swarm_size = 10, inertia = 0.7, local_weight =2 and global weight = 2. We can see that the learning process follows the discrete–continuous search space as expected. It keeps the variable K in a discrete space while handling the alpha variable in a continuous search space. The best position for each iteration is circled in the plot with the iteration number. The spectrum of colors is used to distinguish between particles of each iteration.

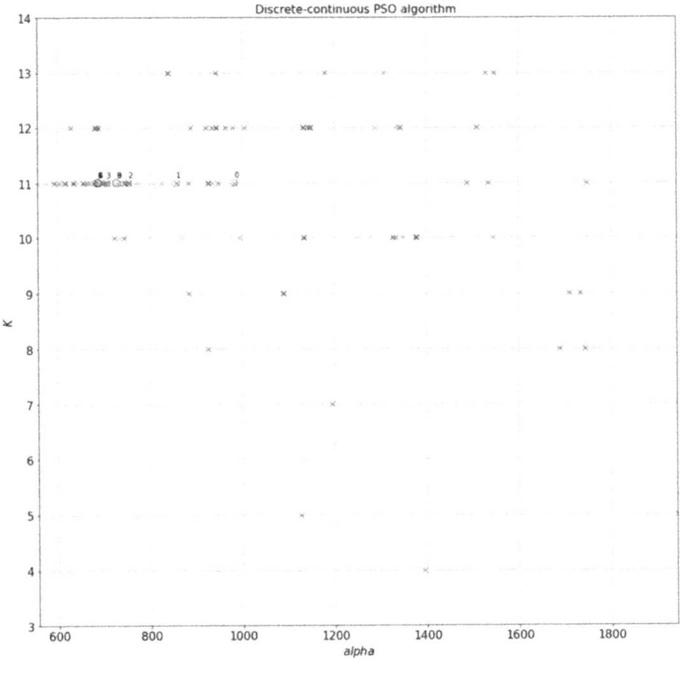


**Figure 3.** Convergence of discrete–continuous PSO algorithm.

**Table 1.** Experiment setup.

| Experiment | Dataset | Description | Referred Literature |
|---|---|---|---|
| 1 | A | Spot price, June 2016, NSW | Wang et al. (2017), Yang et al. (2020) |
| 2 | B<br>C<br>D | Load demand, January 2013, NSW<br>Load demand, April 2013, NSW<br>Load demand, July 2013, NSW | Qiu et al. (2017) |
| 3 | E | Spot price, May 2013, NSW | Peng et al. (2018), Babu et al. (2014) |

Further visualization of the PSO learning process with respect to the fitness value is shown in Figure 4. On the left is the contour plot for the scattered data and on the right is the surface plot of the contour plot. The convergence of the PSO to a global optimum mainly depends on its parameters. The $\beta \times K$ term in the fitness function prevents looking at higher K values in the search space. Thus above-mentioned parameter configuration manages to find near-optimal components for VMD in 10–15 min of time.

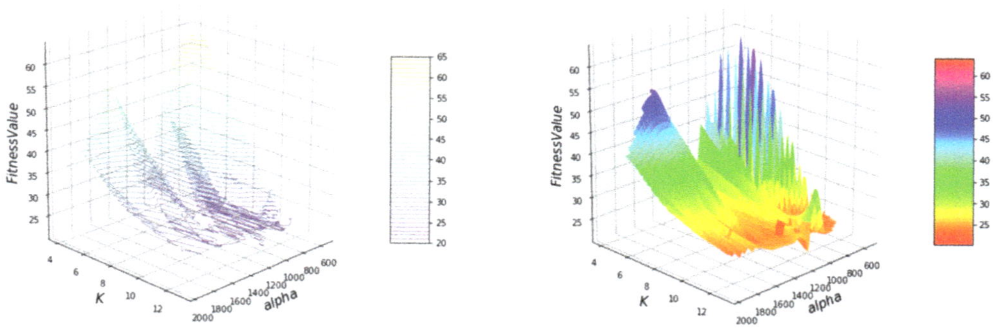


**Figure 4.** D visualization of the PSO learning process.

2.2.4. Time Series Forecasting Module

The forecasting module generates predictions for each sub-series of the input time series data that are decomposed by the VMD algorithm. In the context of predicting sub-series of decomposed input data, each time-step is remodeled; thus, it is not possible to use the previously trained predictive model to predict future values. Therefore, for each new time-step, the predictive model needs to be remodeled, and the re-training process should be efficient and effective to provide an accurate predictive model in a limited amount of time. This duration should ideally be less than the time between two time-steps in the time-series function.

In general, most recent approaches utilize feedforward neural networks; however such feedforward connectionist networks are comparatively slow in training. This slow learning of feedforward neural networks continues to be a major shortcoming for EPF. The key reasons for this latency are the utilization of slow gradient-based learning algorithms and iterative tuning of all parameters of the network during the learning process. In general randomly connected neural networks and Random Vector Functional Link (RVFL) [47] in particular are popular alternative methods for overcoming this limitation. These networks are characterized by the simplicity of RVFL's design and training process. It makes them a very attractive alternative for solving practical machine learning problems in edge computing. Further, our recent result on the efficient FPGA implementation of RVFL [48] makes this type of network particularly suitable for the target real-time prediction scenario

Here we use a variant of RVFL known as Extreme Learning Machines (ELM) [49]. ELM is a single hidden layer feedforward neural network (SLFN) that randomly chooses input weights and analytically determines the output weights. The technical details of the ELM algorithm used for the RIPPR approach are described below.

For $N$ arbitrary distinct input samples $(x_i, t_i)$, where $x_i = [x_{i1}, x_{i2}, \ldots, x_{in}]^T \in R^n$ and $t_i = [t_{i1}, t_{i2}, \ldots, t_{im}]^T \in R^m$ standard SLFNs with $N$ hidden nodes and activation function $g(x)$ are mathematically modelled as:

$$\sum_{i=1}^{\tilde{N}} \beta_i g_i(x_j) = \sum_{i=1}^{\tilde{N}} \beta_i g(w_i.x_j + b_i) = t_j \tag{9}$$

$$j = 1, \ldots\ldots, N$$

where $w_i = [w_{i1}, w_{i2}, \ldots, w_{in}]^T$ is the weight connecting the $i$th hidden node and the input nodes, $\beta_i = [\beta_{i1}, \beta_{i2}, \ldots, \beta_{in}]^T$ is the weight connecting the $i$th hidden node and the output nodes, $\tilde{N}$ is the number of hidden layer nodes, and $b_i$ is the threshold of the $i$th hidden nodes. $w_i \cdot x_i$ denotes the inner product of $w_i$ and $x_i$. The above $N$ equations can be written compactly as:

$$\begin{gathered} \boldsymbol{H}\beta = \boldsymbol{T},\ where \\ \boldsymbol{H}(w_1, \ldots, w_{\tilde{N}}, b_1, \ldots, b_{\tilde{N}}, x_1, \ldots, x_N) = \begin{bmatrix} g(w_1.x_1 + b_1) & \ldots & g(w_{\tilde{N}}.x_1 + b_{\tilde{N}}) \\ \vdots & \ldots & \vdots \\ g(w_1.x_N + b_1) & \ldots & g(w_{\tilde{N}}.x_N + b_{\tilde{N}}) \end{bmatrix}_{N \times \tilde{N}} \\ \beta = \begin{bmatrix} \beta_1^T \\ \vdots \\ \beta_{\tilde{N}}^T \end{bmatrix}_{\tilde{N} \times m} \quad \text{And } \boldsymbol{T} = \begin{bmatrix} t_1^T \\ \vdots \\ t_{\tilde{N}}^T \end{bmatrix}_{\tilde{N} \times m} \end{gathered} \tag{10}$$

where $\boldsymbol{H}$ denotes the hidden layer's output matrix. ELM tends to reach not only the smallest training error but also the smallest norm of output weights. According to Bartlett's theory for feedforward neural networks reaching smaller training error, the smaller the norms of weights are, the better generalization performance of the network.

In the following formulations, 11–15, we deliberate the workings of the learning and generalization of the ELM model. Firstly, output weight optimization is solved as a minimization problem using the generalized inverse matrix of the hidden layer, followed by fine-tuning of the ELM generalization across two cases for N >> L and N > L.

The output weight can be obtained by solving the following minimization problem:

$$Minimize\ : ||\boldsymbol{H}\beta - \boldsymbol{T}||_2\ and\ ||\beta|| \tag{11}$$

where $H$, $\beta$ and $T$ are defined in (10). The reason to minimize the norm of the output weights $||\beta||$ is to maximize the distance of the separating margins of the two different classes in the RVLF feature space.

The optimal solution is given by:

$$\beta = H^{\dagger}\boldsymbol{T} \tag{12}$$

where $H^{\dagger}$ denotes the Moore–Penrose generalized inverse matrix of the hidden layer's output matrix, which can be calculated by the following mathematical transformation. This eliminates the lengthy training phase where network parameters will be adjusted with some hyperparameters in most learning algorithms:

$$H^{\dagger} = \left[H^T H\right]^{-1} H^T \tag{13}$$

Input weights of the SLFN are randomly chosen, then the output weights (linking the hidden layer to the output layer) of an SLFN are analytically determined by the minimum norm least-squares solutions of a general system of linear equations. The running speed of ELM can be a thousand times faster than traditional iterative implementations of SLFNs. To further extend the generalizability of ELM, regularized extreme learning machine algorithm is introduced [50]. The original algorithm is extended by adding a regularization parameter (C) to control the generalization. This is divided into two cases as follows;

Case 1:

If the number of training data is very large, for example, it is much larger than the dimensionality of the feature space,

N >> L:

$$\beta = \left(\frac{I}{C} + H^T H\right)^{-1} H^T T \tag{14}$$

Case 2:

N > L:

$$\beta = H^T \left(\frac{I}{C} + H H^T\right)^{-1} T \tag{15}$$

where $I$ is the identity matrix.

## 3. Experiments and Results

In this section, we evaluate RIPPR on three experiments conducted on five different datasets of EPF for the state of New South Wales (NSW), Australia. The datasets were chosen to reflect the factors of different seasons in Australia. The following section describes the experiments, their datasets and their characteristics.

The experiments were carried out on a multi-core CPU at 2.8 GHz with 16 GB memory and GPU of NVIDIA GeForce GTX 1060.

### *3.1. Experimental Process*

First, we will consider the real-world scenario and then modify it to the experimental study (past data). Here the forecasting horizon is $h$ (i.e., forecasts are generated for $h$ step ahead). The full process is outlined in Algorithm 3.

**Algorithm 3** Experiment procedure

**Input:** Data (X),h,(K,$\alpha$) pair for the given h (taken from the optimization module)
**Output:** h step ahead forecasted value
1: Obtain the most recent 1440 data points from X(1 month period if the data rate is 30 min$^{-1}$);
2: Decompose the data into K sub-series by using the data decomposition module;
3: Divide each sequence (sub-series) into multiple input/output patterns for the given forecasting horizon. Here, we will have (1440-h-input size) samples that have target values (outputs). For the experiment, the input size is kept as 24. We have (1416-h) samples. Call this train set. Last (h) samples will not have a target value. Call this test set;
4: Train on the train data using the time series forecasting module for each sub-series;
5: Predict for the test data using trained models for each sub-series;
6: Aggregate the predicted values for each sub-series to obtain the final prediction for the test data (from the h number of predicted values, the last value will give the final h-step ahead prediction for the given time frame);
7: At the arrival of a new data point, add it to the data set and remove the least recent data point from the data set and go to step (2).

For the experimental study, we start the above procedure starting from the train set and continue till the whole test set values are predicted.

### 3.2. Results

We report the empirical evaluation of RIPPR in terms of the following performance metrics, mean absolute error (MAE), root mean square error (RMSE), mean absolute percentage error (MAPE) and mean squared error (MSE).

#### 3.2.1. Experiment 1

This experiment was designed as a comparative study of results for dataset A, compared between the modules of RIPPR and the available literature [29,30]. The RIPPR modules consist of ELM, VMD-ELM with a fixed K = 8 and $\alpha$ = 1500, VMD-PSO-ELM (Proposed RIPPR approach). The dataset is divided into train and test as follows, as the training dataset first 25 days is used. Therefore, the training dataset consists of 1200 data points. As the test dataset, the last 5 days are used. Thus, it contains 240 data points. The experiment results are shown in Table 2. The experiment compares results in three metrics, namely MAE($/MWh), RMSE($/MWh) and MAPE(%). In the experiment results, in three instances MAE and MAPE are superior to results reported in [28].

**Table 2.** Results comparison for dataset A.

| Horizon (h) | Metric | Persistence | LSTM | ELM | VMD-ELM | RIPPR | Wang et al. [29] | Yang et al. [30] |
|---|---|---|---|---|---|---|---|---|
| 0.5 | MAE | 35.75 | 27.08 | 38.62 | 7.92 | **4.67** | 7.17 | 5.05 |
| | RMSE | 56.38 | 39.76 | 55.32 | 9.26 | **5.63** | 9.77 | 6.61 |
| | MAPE | 31.43 | 23.95 | 30.45 | 8.64 | **4.89** | 7.88 | 6.22 |
| 1 | MAE | 48.48 | 43.40 | 45.11 | 11.94 | 5.45 | 10.54 | **5.04** |
| | RMSE | 71.38 | 60.69 | 62.72 | 14.78 | **6.91** | 14.50 | 7.11 |
| | MAPE | 46.85 | 38.49 | 43.01 | 12.75 | 6.02 | 12.17 | **5.95** |
| 2 | MAE | 66.84 | 63.45 | 56.23 | 13.90 | **9.87** | 15.43 | 10.98 |
| | RMSE | 94.00 | 85.77 | 78.45 | 17.66 | **12.35** | 20.25 | 14.25 |
| | MAPE | 78.41 | 76.45 | 51.26 | 14.05 | **10.21** | 17.64 | 12.94 |
| 3 | MAE | 83.79 | 80.44 | 62.98 | 18.37 | 18.39 | 21.10 | **18.02** |
| | RMSE | 110.87 | 100.99 | 84.64 | 23.69 | **20.01** | 26.61 | 22.52 |
| | MAPE | 108.17 | 81.45 | 63.48 | 18.34 | **18.75** | 24.89 | 21.47 |

Across all instances of this experiment, RIPPR reports a better RMSE value than the literature. A key challenge in EPF is the inability to forecast outliers. From these three metrics, RMSE is the most sensitive metric to outliers. Therefore, we can confirm that our model has a more effective capability to forecast outliers than those reported in the related literature. Optimal component selection of VMD using PSO gained an advantage over the other models. A single step in this experiment represents 30 min of time.

In this comparison (Figures 5–8), we compared five models for dataset A. The models include the Persistence model, LSTM (with two hidden layers), ELM, VMD-ELM (with a constant $\alpha$-1500 and K = 8) and finally RIPPR, which uses PSO to find the optimal components for the VMD algorithm. In the first scenario (1 step ahead forecasting), it is seen that as expected, VMD-ELM outperforms the Persistence model, LSTM and the traditional ELM model by a considerable margin. The capability of RIPPR over VMD-ELM is clearly visible in the second9 scenario (Six steps ahead forecasting), where we can see that the residuals of the RIPPR are significantly lower than the VMD-ELM's residuals. These results confirm that RIPPR can significantly outperform the VMD-ELM model. Due to the lower performance of the Persistence model and the LSTM model, we have excluded them from the later experiments.

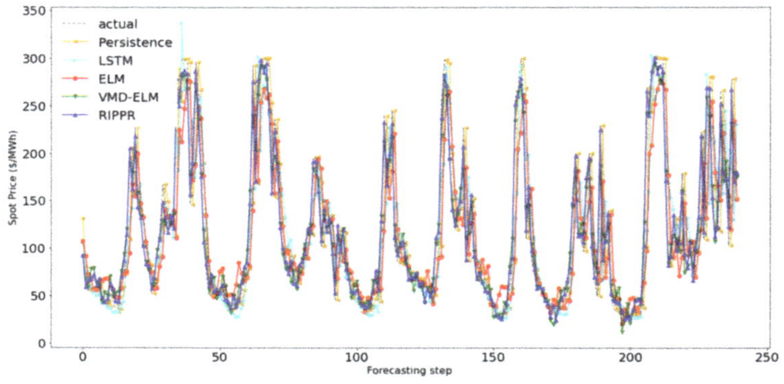


**Figure 5.** One step ahead forecasting for dataset A.

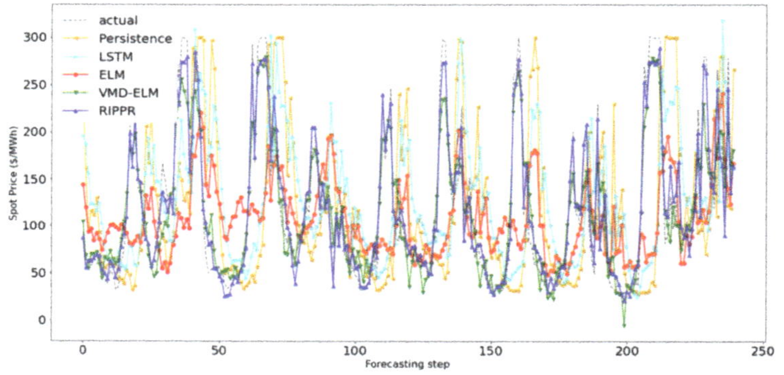


**Figure 6.** Six steps ahead forecasting for dataset A.

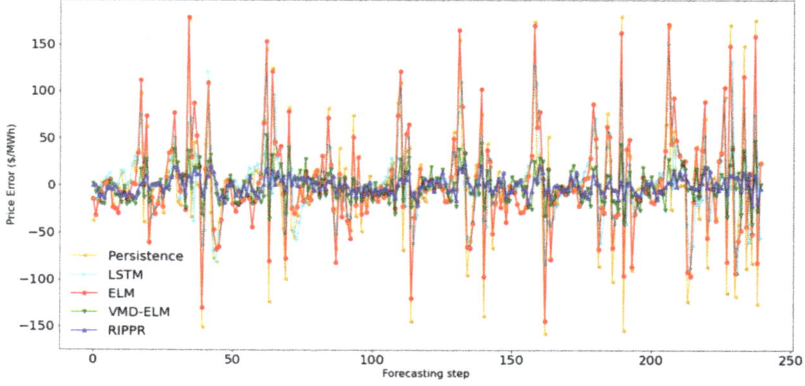


**Figure 7.** Forecasting error-one step ahead forecasting for dataset A.

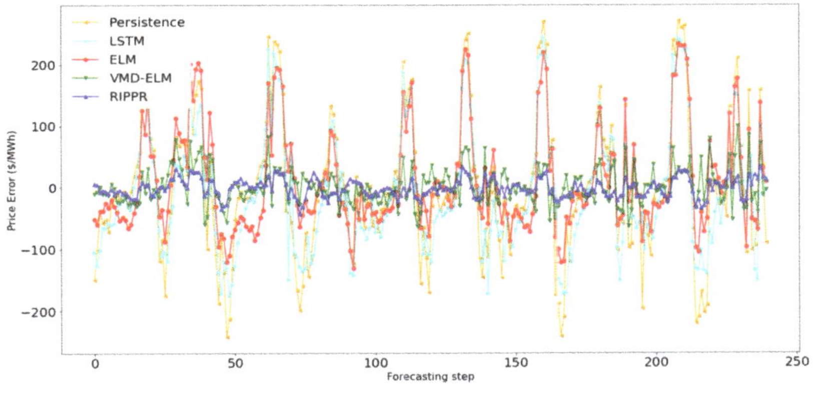


**Figure 8.** Forecasting error-six steps ahead forecasting for dataset A.

Furthermore, to verify the significance of the accuracy improvement of the RIPPR model, the forecasting accuracy comparison with the aforementioned models is conducted using Wilcoxon signed-rank test. It is conducted under a significance level of 0.05 in one-tail-tests. The test results are presented in Table 3. It is clearly seen that there is a statistical significance (under a significance level of 0.05) for the proposed RIPPR among the compared models, including the Persistence model, LSTM model, ELM model and VMD-ELM model.

**Table 3.** Wilcoxon signed-rank test.

| **Compared Models** | **Wilcoxon Signed-Rank Test** | |
|---|---|---|
| | **OneStep Ahead ($\alpha$ = 0.05; W = 1611)** | **SixStep Ahead ($\alpha$ = 0.05; W = 9882)** |
| RIPPR vs. Persistence | 1120 | 608 |
| RIPPR vs. LSTM | 869 | 499 |
| RIPPR vs. ELM | 316 | 734 |
| RIPPR vs. VMD-ELM | 1548 | 6782 |

### 3.2.2. Experiment 2

This experiment was also designed as a comparative study for datasets B, C and D between RIPPR modules as experiment 1 and the available literature [51]. Note that here we consider the electricity load demand for the given time period. The first 3 weeks of each dataset are used to train the model, and the remaining week is used as the test set. Therefore, the training set consists of 1008 data points, and the test set consists of 336 data points. The experiment results are shown in Table 4. The experiment compares results for two metrics, namely RMSE (MW) and MAPE (%). The results clearly indicate that the RIPPR model has outperformed the available literature for all datasets. We can confirm the superiority of VMD over EMD in an EPF scenario as presented in this experiment.

**Table 4.** Results comparison for datasets B, C and D.

| | Horizon (h) | Metric | ELM | VMD-ELM | RIPPR | Qiu et al. |
|---|---|---|---|---|---|---|
| B | 0.5 | RMSE | 94.09 | 37.82 | **28.21** | 49.86 |
| | | MAPE | 0.86 | 0.35 | **0.27** | 0.53 |
| | 24 | RMSE | 754.18 | 483.12 | **420.19** | 541.53 |
| | | MAPE | 6.94 | 4.32 | **4.01** | 4.62 |
| C | 0.5 | RMSE | 115.61 | 46.33 | **37.56** | 69.55 |
| | | MAPE | 1.09 | 0.46 | **0.35** | 0.65 |
| | 24 | RMSE | 567.33 | 400.16 | **352.23** | 377.63 |
| | | MAPE | 5.67 | 3.78 | **2.89** | 3.22 |
| D | 0.5 | RMSE | 142.61 | 37.47 | **30.96** | 75.09 |
| | | MAPE | 1.21 | 0.31 | **0.25** | 0.70 |
| | 24 | RMSE | 583.75 | 375.59 | **318.15** | 322.04 |
| | | MAPE | 4.51 | 2.66 | **2.39** | 3.08 |

3.2.3. Experiment 3

We follow the same configuration as the two previous experiments for dataset E; RIPPR vs. the available literature [52,53]. All the data were converted into hourly data similar to the literature. Thus, 1 day has 24 data points. In total, 744 data points were obtained, and 24 data points were set as test data for one step (one hour) ahead forecasting scenario. For one day (25 steps) ahead forecasting scenario, 168 data points were considered as the test data. The experimental results are shown in Table 5. The experiment compares results in 2 metrics, namely MAE($/MWh) and MSE($/MWh). In the results, the RIPPR model outperforms the compared literature by a considerable margin across all instances. The superiority of a decomposition-based hybrid model over a traditional model is also confirmed by these results. Hour-ahead forecasting is illustrated in Figure 9, and the 24-h ahead forecasting scenario is presented in Figure 10. For the 24-h ahead scenario in Figure 10, the RIPPR model has managed to capture a number of outliers in the dataset. Further, it is supported by the low MSE values across the two horizons. A single step in this experiment represents one hour of time.

**Table 5.** Results comparison for dataset E.

| Horizon (h) | Metric | ELM | VMD-ELM | RIPPR | Peng et al. [52] | Babu et al. [53] |
|---|---|---|---|---|---|---|
| 1 | MAE | 2.95 | 2.91 | **2.52** | 3.19 | 3.23 |
| | MSE | 13.53 | 13.02 | **10.22** | 15.44 | 18.27 |
| 24 | MAE | 7.85 | 7.48 | **4.63** | 5.01 | 5.32 |
| | MSE | 119.01 | 112.09 | **50.08** | 52.59 | 53.00 |

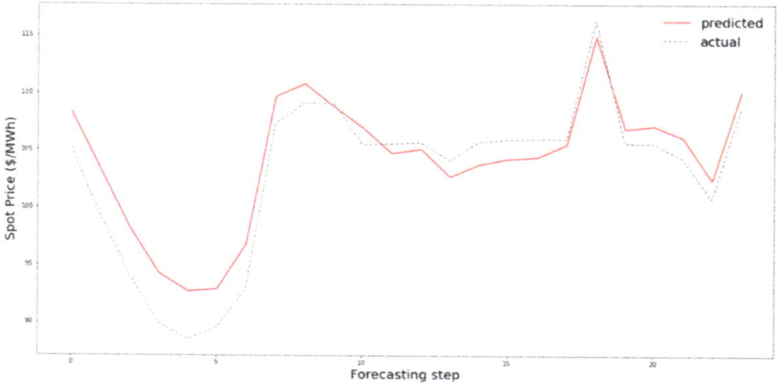


**Figure 9.** Forecasting performance of RIPPR for one hour ahead.

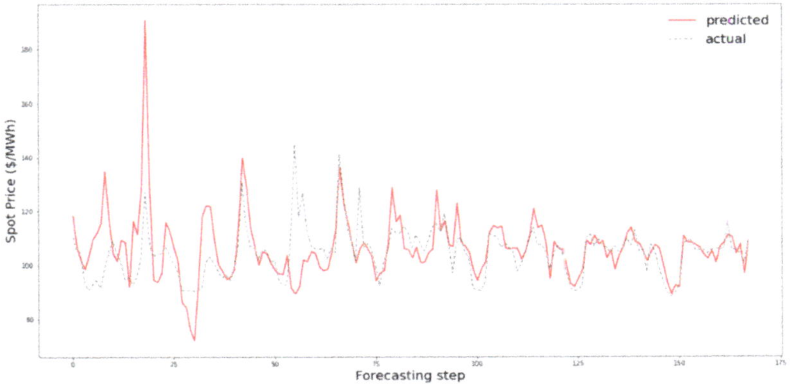


**Figure 10.** Forecasting performance of RIPPR for 24 h ahead.

## 4. Discussion and Conclusions

In this paper, we propose a novel Artificial Intelligence (AI) based approach for electricity price forecasting that addresses the challenges of accuracy, robustness and real-time multi-step prediction. RIPPR utilizes Variational Mode Decomposition (VMD) to transform the spot price data stream into sub-series that are optimized for robustness using particle swarm optimization (PSO). These sub-series are input to an Extreme Learning Machine (ELM) algorithm for real-time multi-step prediction. RIPPR was evaluated with six electricity price/load demand datasets from the Australian energy market. Five benchmark methods were compared with the proposed model to verify its effectiveness. Based on this robust empirical evaluation across three data streams from different market types, we can conclude that VMD based hybrid models outperform traditional single structure models in EPF, the performance of VMD depends on the mode number (k) and balancing parameter (α), and PSO optimization to find the optimal (k, α) combination improves the results significantly rather than using a static (k, α) combination. As future work, we intend to extend the proposed model to incorporate additional features such as weather, global market variables and related external events that will improve the forecast accuracy and contribute towards the AI capability for real-time monitoring of future smart grids.

**Author Contributions:** Conceptualization, S.K., D.D.S., S.S., R.N., A.J. and V.V.; formal analysis, D.A. and V.V.; investigation, S.K. and A.J.; methodology, S.K., D.D.S., S.S., R.N., E.O. and A.J.; resources, D.A.; software, V.V.; validation, D.D.S., S.S., R.N. and E.O.; writing—original draft, S.K., D.D.S. and S.S.; writing—review and editing, D.A., E.O., A.J. and V.V. All authors have read and agreed to the published version of the manuscript.

**Funding:** This research received no external funding.

**Institutional Review Board Statement:** Not applicable.

**Informed Consent Statement:** Not applicable.

**Data Availability Statement:** All data is publicly available from the Australian Energy Market Operator (AEMO) at https://aemo.com.au/ accessed on 4 May 2021.

**Conflicts of Interest:** The authors declare no conflict of interest.

## Nomenclature

| Nomenclature | Refferd to |
|---|---|
| VMD | Variational Mode Decomposition |
| PSO | particle swarm optimization |

| | |
|---|---|
| RVFL | Random Vector Functional Link neural network |
| ELM | Extreme Learning Machine |
| EPF | Electircity Price Forecasting |
| ARMA | Auto regressive moving average |
| ARIMA | Auto-regressive integrated moving average |
| VAR | Vector auto-regression |
| GARCH | Generalized autoregressive conditional heteroskedasticity |
| ANN | Artificial neural networks |
| SVM | Support vector machine |
| FNN | Fuzzy neural networks |
| RNN | Recurrent neural networks |
| ERCOT | Electric Reliability Council of Texas |
| LSTM | Long short-term memory |
| WPT | Wavelet packet transform |
| PSOSA | particle swarm optimization based on simulated annealing |
| LSSVM | Least Square Support Vector Machine |
| FEEMD | Fast ensemble empirical mode decomposition |
| BP | Back propagation |
| IMOSCA | Improved multi-objective sine cosine algorithm |
| RELM | Regularized extreme learning machine |
| EMD | Empirical Mode decomposition |
| EWMA | Exponential Weighted Moving Average |
| EWT | Empirical Wavelet Transform |
| SLFN | Single hidden layer feedforward neural network |
| NSW | New South Wales |
| MAE | Mean absolute error |
| RMSE | Root mean square error |
| MAPE | Mean absolute percentage error |

## References


1. Li, Z.; Qiu, F.; Wang, J. Data-driven real-time power dispatch for maximizing variable renewable generation. *Appl. Energy* **2016**, *170*, 304–313. [CrossRef]
2. Petrollese, M.; Valverde-Isorna, L.; Cocco, D.; Cau, G.; Guerra, J. Real-time integration of optimal generation scheduling with MPC for the energy management of a renewable hydrogen-based microgrid. *Appl. Energy* **2016**, *166*, 96–106. [CrossRef]
3. Li, Y.; Wang, C.; Li, G.; Chen, C. Optimal scheduling of integrated demand response-enabled integrated energy systems with uncertain renewable generations: A Stackelberg game approach. *Energy Convers. Manag.* **2021**, *235*, 113996. [CrossRef]
4. Kühnlenz, F.; Nardelli, P.H.; Karhinen, S.; Svento, R. Implementing flexible demand: Real-time price vs. market integration. *Energy* **2018**, *149*, 550–565. [CrossRef]
5. Lujano-Rojas, J.M.; Zubi, G.; Dufo-López, R.; Bernal-Agustín, J.L.; Catalão, J.P. Novel probabilistic optimization model for lead-acid and vanadium redox flow batteries under real-time pricing programs. *Int. J. Electr. Power Energy Syst.* **2018**, *97*, 72–84. [CrossRef]
6. Barhagh, S.S.; Abapour, M.; Mohammadi-Ivatloo, B. Optimal scheduling of electric vehicles and photovoltaic systems in residential complexes under real-time pricing mechanism. *J. Clean. Prod.* **2020**, *246*, 119041. [CrossRef]
7. Jiang, J.; Kou, Y.; Bie, Z.; Li, G. Optimal Real-Time Pricing of Electricity Based on Demand Response. *Energy Procedia* **2019**, *159*, 304–308. [CrossRef]
8. Wang, F.; Ge, X.; Yang, P.; Li, K.; Mi, Z.; Siano, P.; Duić, N. Day-ahead optimal bidding and scheduling strategies for DER aggregator considering responsive uncertainty under real-time pricing. *Energy* **2020**, *213*, 118765. [CrossRef]
9. Anees, A.; Dillon, T.; Wallis, S.; Chen, Y.-P.P. Optimization of day-ahead and real-time prices for smart home community. *Int. J. Electr. Power Energy Syst.* **2021**, *124*, 106403. [CrossRef]
10. Anand, H.; Ramasubbu, R. A real time pricing strategy for remote micro-grid with economic emission dispatch and stochastic renewable energy sources. *Renew. Energy* **2018**, *127*, 779–789. [CrossRef]
11. Xu, B.; Wang, J.; Guo, M.; Lu, J.; Li, G.; Han, L. A hybrid demand response mechanism based on real-time incentive and real-time pricing. *Energy* **2021**, *231*, 120940. [CrossRef]
12. Botelho, D.; Dias, B.; de Oliveira, L.; Soares, T.; Rezende, I.; Sousa, T. Innovative business models as drivers for prosumers integration-Enablers and barriers. *Renew. Sustain. Energy Rev.* **2021**, *144*, 111057. [CrossRef]
13. Elma, O.; Taşcıkaraoğlu, A.; Ince, A.T.; Selamogullari, U.S. Implementation of a dynamic energy management system using real time pricing and local renewable energy generation forecasts. *Energy* **2017**, *134*, 206–220. [CrossRef]
14. Mbungu, N.T.; Bansal, R.C.; Naidoo, R.; Miranda, V.; Bipath, M. An optimal energy management system for a commercial building with renewable energy generation under real-time electricity prices. *Sustain. Cities Soc.* **2018**, *41*, 392–404. [CrossRef]

15. Mirakhorli, A.; Dong, B. Market and behavior driven predictive energy management for residential buildings. *Sustain. Cities Soc.* **2018**, *38*, 723–735. [CrossRef]
16. Finck, C.; Li, R.; Zeiler, W. Optimal control of demand flexibility under real-time pricing for heating systems in buildings: A real-life demonstration. *Appl. Energy* **2020**, *263*, 114671. [CrossRef]
17. Forrest, S.; MacGill, I. Assessing the impact of wind generation on wholesale prices and generator dispatch in the Australian National Electricity Market. *Energy Policy* **2013**, *59*, 120–132. [CrossRef]
18. Alahyari, A.; Pozo, D. Electric end-user consumer profit maximization: An online approach. *Int. J. Electr. Power Energy Syst.* **2021**, *125*, 106502. [CrossRef]
19. Thilker, C.A.; Madsen, H.; Jørgensen, J.B. Advanced forecasting and disturbance modelling for model predictive control of smart energy systems. *Appl. Energy* **2021**, *292*, 116889. [CrossRef]
20. Chatfield, C.; Routledge, B. *Time-Series Forecasting*, 1st ed.; CRC Press: Boca Raton, FL, USA, 2000; Available online: https://www.routledge.com/Time-Series-Forecasting/Chatfield/p/book/9781584880639 (accessed on 14 February 2021).
21. Chujai, P.; Kerdprasop, N.; Kerdprasop, K. Time series analysis of household electric consumption with ARIMA and ARMA models. *Lect. Notes Eng. Comput. Sci.* **2013**, *2202*, 295–300.
22. García-Ascanio, C.; Maté, C. Electric power demand forecasting using interval time series: A comparison between VAR and iMLP. *Energy Policy* **2010**, *38*, 715–725. [CrossRef]
23. Girish, G. Spot electricity price forecasting in Indian electricity market using autoregressive-GARCH models. *Energy Strat. Rev.* **2016**, *11–12*, 52–57. [CrossRef]
24. Ma, Z.; Zhong, H.; Xie, L.; Xia, Q.; Kang, C. Month ahead average daily electricity price profile forecasting based on a hybrid nonlinear regression and SVM model: An ERCOT case study. *J. Mod. Power Syst. Clean Energy* **2018**, *6*, 281–291. [CrossRef]
25. Anand, A.; Suganthi, L. Forecasting of Electricity Demand by Hybrid ANN-PSO Models. In *Deep Learning and Neural Networks*; IGI Global: Hershey, PA, USA, 2020; pp. 865–882. Available online: https://www.igi-global.com/gateway/chapter/237910 (accessed on 3 March 2021).
26. Yunpeng, L.; Di, H.; Junpeng, B.; Yong, Q. Multi-step Ahead Time Series Forecasting for Different Data Patterns Based on LSTM Recurrent Neural Network. In Proceedings of the 2017 14th Web Information Systems and Applications Conference (WISA), Liuzhou, China, 11–12 November 2017; pp. 305–310.
27. Hassan, S.; Khosravi, A.; Jaafar, J.; Khanesar, M.A. A systematic design of interval type-2 fuzzy logic system using extreme learning machine for electricity load demand forecasting. *Int. J. Electr. Power Energy Syst.* **2016**, *82*, 1–10. [CrossRef]
28. Wang, J.-Z.; Wang, Y.; Jiang, P. The study and application of a novel hybrid forecasting model–A case study of wind speed forecasting in China. *Appl. Energy* **2015**, *143*, 472–488. [CrossRef]
29. Wang, D.; Luo, H.; Grunder, O.; Lin, Y.; Guo, H. Multi-step ahead electricity price forecasting using a hybrid model based on two-layer decomposition technique and BP neural network optimized by firefly algorithm. *Appl. Energy* **2017**, *190*, 390–407. [CrossRef]
30. Yang, W.; Wang, J.; Niu, T.; Du, P. A novel system for multi-step electricity price forecasting for electricity market management. *Appl. Soft Comput.* **2020**, *88*, 106029. [CrossRef]
31. He, K.; Wang, H.; Du, J.; Zou, Y. Forecasting Electricity Market Risk Using Empirical Mode Decomposition (EMD)—Based Multiscale Methodology. *Energies* **2016**, *9*, 931. [CrossRef]
32. Liu, Z.; Jiang, P.; Zhang, L.; Niu, X. A combined forecasting model for time series: Application to short-term wind speed forecasting. *Appl. Energy* **2020**, *259*, 114137. [CrossRef]
33. Abdollahi, H. A novel hybrid model for forecasting crude oil price based on time series decomposition. *Appl. Energy* **2020**, *267*, 115035. [CrossRef]
34. Chitalia, G.; Pipattanasomporn, M.; Garg, V.; Rahman, S. Robust short-term electrical load forecasting framework for commercial buildings using deep recurrent neural networks. *Appl. Energy* **2020**, *278*, 115410. [CrossRef]
35. Somu, N.; MR, G.R.; Ramamritham, K. A hybrid model for building energy consumption forecasting using long short term memory networks. *Appl. Energy* **2020**, *261*, 114131. [CrossRef]
36. Jallal, M.A.; González-Vidal, A.; Skarmeta, A.F.; Chabaa, S.; Zeroual, A. A hybrid neuro-fuzzy inference system-based algorithm for time series forecasting applied to energy consumption prediction. *Appl. Energy* **2020**, *268*, 114977. [CrossRef]
37. Huang, N.E.; Shen, Z.; Long, S.R.; Wu, M.C.; Shih, H.H.; Zheng, Q.; Yen, N.-C.; Tung, C.C.; Liu, H.H. The empirical mode decomposition and the Hilbert spectrum for nonlinear and non-stationary time series analysis. *Proc. R. Soc. A Math. Phys. Eng. Sci.* **1998**, *454*, 903–995. [CrossRef]
38. Wu, Z.H.; Huang, N.E. Ensemble empirical mode decomposition: A noise-assistant data analysis method. *Adv. Adapt. Data Anal.* **2009**, *1*, 1–41. [CrossRef]
39. Torres, M.E.; Colominas, M.A.; Schlotthauer, G.; Flandrin, P. A complete ensemble empirical mode decomposition with adaptive noise. In Proceedings of the 2011 IEEE International Conference on Acoustics, Speech and Signal Processing (ICASSP), Prague, Czech Republic, 22–27 May 2011; pp. 4144–4147. [CrossRef]
40. Gilles, J. Empirical Wavelet Transform. *IEEE Trans. Signal. Process.* **2013**, *61*, 3999–4010. [CrossRef]
41. Dragomiretskiy, K.; Zosso, D. Variational Mode Decomposition. *IEEE Trans. Signal. Process.* **2014**, *62*, 531–544. [CrossRef]
42. Wang, Y.; Markert, R.; Xiang, J.; Zheng, W. Research on variational mode decomposition and its application in detecting rub-impact fault of the rotor system. *Mech. Syst. Signal. Process.* **2015**, *60–61*, 243–251. [CrossRef]

43. Aneesh, C.; Kumar, S.; Hisham, P.; Soman, K. Performance Comparison of Variational Mode Decomposition over Empirical Wavelet Transform for the Classification of Power Quality Disturbances Using Support Vector Machine. *Procedia Comput. Sci.* **2015**, *46*, 372–380. [CrossRef]
44. Hossain, M.A.; Chakrabortty, R.K.; Ryan, M.; Pota, H.R. Energy Management of Community Microgrids Considering Uncertainty using Particle Swarm Optimisation. *Preprints* **2020**. [CrossRef]
45. Kennedy, J.; Eberhart, R. particle swarm optimization. In Proceedings of the ICNN'95 International Conference on Neural Networks, Perth, Australia, 27 November–1 December 1995; Volume 4, pp. 1942–1948.
46. Prakash, K.N.; Sydulu, M. Particle Swarm Optimization Based Capacitor Placement on Radial Distribution Systems. In Proceedings of the 2007 IEEE Power Engineering Society General Meeting, Tampa, FL, USA, 24–28 June 2007; pp. 1–5.
47. Igelnik, B.; Pao, Y.-H. Stochastic choice of basis functions in adaptive function approximation and the functional-link net. *IEEE Trans. Neural Netw.* **1995**, *6*, 1320–1329. [CrossRef]
48. Kleyko, D.; Kheffache, M.; Frady, E.P.; Wiklund, U.; Osipov, E. Density Encoding Enables Resource-Efficient Randomly Connected Neural Networks. *IEEE Trans. Neural Networks Learn. Syst.* **2020**, 1–7. [CrossRef]
49. Huang, G.B.; Zhu, Q.Y.; Siew, C.K. Extreme Learning Machine: A New Learning Scheme of Feedforward Neural Networks. In Proceedings of the 2004 IEEE International Joint Conference on Neural Networks—Proceedings, Budapest, Hungary, 25–29 July 2004.
50. Huang, G.-B.; Zhou, H.; Ding, X.; Zhang, R. Extreme Learning Machine for Regression and Multiclass Classification. *IEEE Trans. Syst. Man Cybern. Part B* **2012**, *42*, 513–529. [CrossRef] [PubMed]
51. Qiu, X.; Ren, Y.; Suganthan, P.; Amaratunga, G.A. Empirical Mode Decomposition based ensemble deep learning for load demand time series forecasting. *Appl. Soft Comput.* **2017**, *54*, 246–255. [CrossRef]
52. Peng, L.; Liu, S.; Liu, R.; Wang, L. Effective long short-term memory with differential evolution algorithm for electricity price prediction. *Energy* **2018**, *162*, 1301–1314. [CrossRef]
53. Babu, C.N.; Reddy, B.E. A moving-average filter based hybrid ARIMA–ANN model for forecasting time series data. *Appl. Soft Comput.* **2014**, *23*, 27–38. [CrossRef]

energies

Article

# Intelligent Energy Management in a Prosumer Community Considering the Load Factor Enhancement

Fernando V. Cerna [1], Mahdi Pourakbari-Kasmaei [2,*], Luizalba S. S. Pinheiro [1], Ehsan Naderi [3], Matti Lehtonen [2] and Javier Contreras [4]

1 Department of Electrical Engineering, Federal University of Roraima, Boa Vista 69310-000, Brazil; Fernando.Cerna@ufrr.br (F.V.C.); Luizalba.Pinheiro@ufrr.br (L.S.S.P.)
2 Department of Electrical Engineering and Automation, Aalto University, 02150 Espoo, Finland; Matti.Lehtonen@aalto.fi
3 School of Electrical, Computer, and Biomedical Engineering, Southern Illinois University, Carbondale, IL 62901, USA; Ehsan.Naderi@siu.edu
4 School of Industrial Engineering, University of Castilla-La Mancha, 13071 Ciudad Real, Spain; Javier.Contreras@uclm.es
* Correspondence: Mahdi.Pourakbari@aalto.fi; Tel.: +358-50-446-0840

**Abstract:** In prosumers' communities, the use of storage batteries (SBs) as support for photovoltaic (PV) sources combined with coordination in household appliances usage guarantees several gains. Although these technologies increase the reliability of the electricity supply, the large-scale use of home appliances in periods of lower solar radiation and low electricity tariff can impair the performance of the electrical system. The appearance of new consumption peaks can lead to disturbances. Moreover, the repetition of these events in the short term can cause rapid fatigue of the assets. To address these concerns, this research proposes a mixed-integer linear programming (MILP) model aiming at the optimal operation of the SBs and the appliance usage of each prosumer, as well as a PV plant within a community to achieve the maximum load factor (LF) increase. Constraints related to the household appliances, including the electric vehicle (EV), shared PV plant, and the SBs, are considered. Uncertainties in consumption habits are simulated using a Monte Carlo algorithm. The proposed model was solved using the CPLEX solver. The effectiveness of our proposed model is evaluated with/without the LF improvement. Results corroborate the efficient performance of the proposed tool. Financial benefits are obtained for both prosumers and the energy company.

**Keywords:** community of prosumers; new consumption peak; shared PV plant; storage batteries; load factor



**Citation:** Cerna, F.V.; Pourakbari-Kasmaei, M.; Pinheiro, L.S.S.; Naderi, E.; Lehtonen, M.; Contreras, J. Intelligent Energy Management in a Prosumer Community Considering the Load Factor Enhancement. *Energies* **2021**, *14*, 3624. https://doi.org/10.3390/en14123624

Academic Editor: Luis Hernández-Callejo

Received: 19 May 2021
Accepted: 16 June 2021
Published: 18 June 2021



## 1. Introduction

### 1.1. Overview

The need for efficient electricity management in large cities worldwide has led electric utilities to implement the *Smart Grid* (SG) concept in their electrical distribution networks (EDNs) [1]. In an advanced communication environment, the SG is characterized by the bidirectional flow of data and power between the smart meters of customers with the information center of energy companies [1,2]. In this context, traditional electricity consumers play an active role in the electrical grid operation either through the photovoltaic (PV) generation [3] or the change in habits of usage of residential appliances [4]. In the case of PV generation, some part is used for the customer's own consumption, while the surplus PV energy is injected into the power grid [5,6]. In this way, energy credits are obtained by the prosumers, allowing them to reduce the energy bill for the coming months [6]. Moreover, sometimes, in smart homes, energy storage devices are used to mitigate the intermittency in PV production during the day to ensure supply continuity [7,8]. In the case of a consumption habit change, the appliances whose scheduling is based on hourly rates allow reducing energy consumption during periods of higher energy prices, i.e., peak consumption period [4]. The aspects mentioned above represent the most well-known strategies to guarantee the security and reliability of the EDN in the SG environment [9]. However, the diurnal and intermittent characteristics of solar radiation, as well as the dynamic energy tariffs on a given day, can contribute to the appearance of new consumption peaks in the consumption profile of the community [10,11]. As during off-peak hours the energy tariff is cheaper, many prosumers prefer to postpone or anticipate the usage of the appliance, then the usage of their appliances may coincide, especially those with higher average power, within these periods [12]. These events can affect the power grid performance, compromising the assets lifetime, mainly of the power transformers, feeders and protection devices [13].

The International Energy Agency (IEA), in its 2017 report, *Residential Prosumers in the European Energy Union*, highlights the impact on the electricity grid (e.g., congestion and its stability), focusing on the increase in the number of prosumers in a short term horizon [14]. Upon these challenges, the development of smart tools that assist decision makers toward efficient energy management of prosumers communities is of critical importance. Furthermore, such tools should consider the assessment of indexes related to the rational usage of energy to guarantee the efficiency of the power supply. One way can be through the evaluation of the load factor (LF). The LF is defined as the ratio between the average demand and the maximum demand for a given period [15]. This indicator varies over a range of values, with a minimum of 0 and a maximum of 1 [16]. Depending on the adopted value, the LF related to a given consumption profile indicates the efficiency level at which the electric energy is being utilized. For example, when the LF value is low, e.g., 0.5, 0.4, 0.2, etc., the profile shows high energy consumption (peaks) at different times of the day, as well as the periods at which the consumption is almost zero (valleys). In contrast, when the value is close to 1, e.g., 0.75, 0.8, 0.9, etc., the profile shows a wide distribution of energy consumption throughout the day, indicating efficient electricity management [15,17]. Therefore, the LF improvement as part of the intelligent energy management strategy of prosumers communities can contribute to planning the EDN more economically and efficiently.

The literature review below highlights the existence of a gap related to the topic of energy communities to be addressed. To address such shortcomings and fill the existing gap, this paper proposes a MILP model to manage electricity consumption in a prosumer community efficiently. The proposed model aims to minimize the energy purchased from the power grid by scheduling the loads. At the same time, a system of SBs combined with a shared PV plant fulfills the energy needs of the prosumer community, and, in the case of generating surpluses of power, these are directed to the grid. These objectives are achieved in a coordinated manner to obtain a wide distribution of consumption during the day, which is represented by improving the LF. By increasing the LF, this work seeks to mitigate the appearance of new peaks in periods with cheap energy tariffs (due to the coincident usage of appliances with higher average power) via intelligent operation of PV-based technology and energy storage, ensuring continuity of supply. Operational constraints related to the operation periods of household appliances and battery charging of EVs within each smart home, SBs, shared PV plant, and power balance in the power distribution system are taken into account. Moreover, the uncertainties in the usage patterns of household appliances and charging of EVs are simulated using a Monte Carlo algorithm. Our work seeks to go beyond the model proposed in [18]. In this research, the problem of improving the LF was addressed to a group of consumers differentiated by domestic income, that is, each consumer having a different number of appliances. In addition, in this research, the group of consumers has been supplied solely by the energy company, disregarding the presence of sources of electricity generation and/or storage. This work addresses the empowerment of residential consumers (especially those with higher household income), through the use of power technologies to generate surplus energy once all the community's energy requirements are met. The proposed MILP model is implemented in the AMPL [19] language, and to guarantee to obtain the global optimal solution, the commercial solver CPLEX [20] is used. Consequently, the main contributions of this research are itemized hereunder.

- Proposing a computationally efficient MILP model to improve the value of LF related to the consumption profile of prosumers while taking into account the efficient scheduling of technologies such as SBs and shared PV generation.
- Investigating the intelligent management of an energy community by improving an indicator of the rational usage of energy.
- Establishing the management of household appliances (including the EV) to avoid their coincident consumption, especially those with higher average power, to mitigate the occurrence of peak consumption in off-peak periods and/or with insufficient levels of solar irradiation.
- Contributing to reducing the dependence on fossil fuels to meet the energy of domestic customers aiming at a sustainability context.

### 1.2. Literature Review

Most of the works addressing energy management of prosumers communities focused on trading the purchase/sale of energy among stakeholders. The authors in [21] addressed the problem of energy consumption scheduling for the day-ahead. Cooperative game theory was implemented to model the scheduling strategy aiming at minimizing the total costs of each prosumer. Moreover, this strategy was tested in a community with ten energy self-producers. The impact of consumption patterns of the prosumers, as well as the levels of solar radiation in the integration of the PV generation in the communities, was investigated in [22]. The primary objective was to reduce the total costs related to investments in power technologies (i.e., PV panels and SBs), and results revealed considerable savings for prosumers, mainly in the electricity bill. The authors in [23] proposed an optimization model to represent several regulatory aspects related to tariff schemes and the self-consumption of PV energy. Real data has been used to determine the economic implications of these regulatory mechanisms. The profitability obtained by the consumer community shows its strong dependence on regulatory incentives. A MILP model that aimed to minimize the energy bill of prosumers was developed in [24], where the constraints related to the operation of PV units in each residence were considered. The results highlighted the monetary benefits and the possibility of adapting the contracted demand to the new consumption profile. The work in [25] implemented a two-stage control architecture to efficiently schedule residential loads taking into account thermal comfort. The first stage aimed to allocate the PV power according to the production level of each building. Subsequently, the load of each building was scheduled at the second stage. The results indicated that flexibility in the allocation of power is a key factor for occupant comfort. Similarly, the authors in [26] developed a strategy for controlling the injected PV energy into the power grid. The purpose of this strategy was to stabilize the voltage profile in the distribution system. For this purpose, a non-cooperative game-based algorithm was implemented to control PV generators. In [27] and [28], techniques for optimal management of PV systems, SBs, as well as shared power strategies of energy communities were applied. In [27], an adaptive robust optimization (ARO) structure was used to reduce the consumption costs of prosumers, as well as to optimize the management of thermal loads, while [28] proposed a technique for controlling priority electrical loads together with SBs, both according to flexibility in community consumption. The authors in [29] developed a community energy market model (CEM) aiming at maximizing the financial benefits of both consumers and prosumers. For the consumers, this goal was achieved through the reduction in consumption costs, and for the prosumers by managing the power injected into the electrical network. The work revealed that, depending on the configuration in the consumer and prosumer communities, the economic gains could increase. Another case of CEMs was addressed in [30] by proposing two schemes based on the coordination of generation sources and energy storage within a community. These schemes considered game theory, Karush–Kuhn–Tucker conditions, and strong duality theorem to allow achieving a win-win situation for all participants. To manage the power of a community in real-time, a fog-based model was developed in [31] aiming to improve the energy contract for all electricity self-producers. Three scenarios were presented to evaluate this model, such as (i) supply of demand by the electricity company, (ii) cooperation between prosumers and microgrids to meet demand, and (iii) integration of the electricity company, prosumers, and microgrids. The results show the effectiveness of the fog model for real-time data flow in addition to the financial gain. To minimize the operating costs of both individual residences and an entire residential community, a peer-to-peer (P2P) energy trading methodology was developed in [32]. Each residence was equipped with a rooftop PV system, SBs, and a set of DC and AC electrical loads, where the P2P methodology resulted in energy savings for different levels of PV energy penetration. The authors in [33] implemented two approaches related to prosumers' microgrids. In the first approach, a genetic algorithm was developed to improve the incompatibility between generation and demand of microgrids. The second approach is aimed at optimizing the coordination of SBs charging/discharging. Both approaches were carried out in a regulatory context in Spain and contributed to the efficient performance of the electricity grid. To minimize the total operational costs of a set of prosumers in the EDN, the authors in [34] developed strategies based on the Directed Steiner Tree (DST) and Weighted Dominating Set (WDS) algorithms. In this work, the economic and operational constraints of each microgrid belonging to a given prosumer were considered in formulating the problem. A MILP model was proposed in [35] for optimal sizing of SBs, PV panels, and inverters in the prosumer residence. This model was tested in a deterministic and stochastic manner, demonstrating its potential to establish more economical and operational plans. Some works in the literature on the same topic have used machine learning techniques to better capture different aspects. Suitable adaptation techniques and learning strategies were proposed in [36] to model the behavior patterns of prosumers, as well as the levels of participation in the power grid through power injection. In [37], machine learning strategies were applied to manage the renewable resources of a community of prosumers. In this work, the objective was to minimize the energy consumption expenditures of the community. The aforementioned works highlight the application of strategies based on game theory and artificial intelligence, among others, where most surveys aim to reduce expenses related to energy consumption in the prosumers community by reducing peak demand and/or increasing their financial gain by selling energy to EDN operators. However, there is little attention to research related to the appearance of new consumption peaks in periods with cheaper energy prices by reducing consumption in peak periods through the intelligent management of SBs and accomplishing the shared PV plant.

Furthermore, some works addressed the increase in LF through energy efficiency, integration of distributed generation, or recharging/discharging EVs in the power grid. The authors in [38] implemented a methodology to improve the LF aiming at reducing the active losses, as well as integrating the medium voltage PV plants and plug-in EVs. In [39], the LF improvement through scheduling the charging of EVs was addressed while minimizing the total costs. This work considered two cases. In the first case, through operational planning, the charging of EVs was evaluated. And, in the second case, the LF is improved by charging EVs through efficient scheduling. The LF behavior of a university building was studied in [40] over a period of three years, where strategies related to loading balance were implemented to enhance the building LF. Similarly, in [16], the LF of another university building was improved by replacing old equipment, as well as scheduling priority loads. Through a multi-objective formulation and several stochastic-based cases, the work in [41] aimed to minimize the energy costs of a residential customer and increase the LF related to the domestic consumption profile. The effectiveness of the proposed method was confirmed by obtaining a reduction in customer bills while guaranteeing a satisfactory LF. An operational technique to reduce the peak load was implemented in [17]. In this work, based on classified data according to the frequency of electricity consumption, the household appliances having the most significant impact on the energy bill were identified. By applying this strategy, the peak demand for a set of residential customers was reduced and, consequently, the LF improved To improve the LF of a group of commercial buildings, the authors in [42] proposed demand-side management (DSM) strategy aiming at reducing energy consumption costs. In addition, a tariff structure to minimize peak demand was considered. The results showed a reduction in electricity costs, in addition to an improvement in the performance of the distribution network. As can be seen, in these works, the improvement of the LF was addressed by some surveys that consider the presence of distributed generation and fleets of EVs in the operation of the EDN, while others have considered the implementation of energy efficiency programs in public buildings. However, strategies related to intelligent energy management that consider the LF enhancement as an indicator of efficient electricity usage were disregarded.

The rest of the paper is organized as follows. In Section 2, the main hypotheses that drive the MILP model are presented, while the Monte Carlo simulation algorithm and the SG environment for this problem are explained. Section 3 describes the MILP model in detail. Simulation results and analysis are discussed in Section 4. Finally, Section 5 concludes the paper.

## 2. Simulation Setup

In this Section, the main hypotheses related to the set of household appliances, PV plant, and operation of SBs are presented. The treatment of the hourly preferences of each prosumer, and the simulation algorithm in the habitual usage of household appliances, are explained. Additionally, the flow of energy and information in the SG environment is mentioned in detail.

### 2.1. Hypotheses

The following hypotheses are used to develop the model and analyze its potential are as follows.

1. The research is carried out in the SG environment depicted in Figure 1, which highlights the bidirectional flows between various technologies.
2. Considering that household income is proportional to the number of appliances present at home, it is assumed that all consumers have the same household income taking into account the appliances reported in Tables 1 and 2 including the presence of a single EV (to be charged within each household) according to Table 3.
3. The habitual consumption of each appliance (including the EV) for each period of the day is obtained using the Monte Carlo simulation algorithm.
4. The study horizon considers one day, which is divided into 24 hourly periods.
5. A tariff structure is divided into three levels (peak, intermediate, and off-peak) to efficiently schedule the consumption periods of household appliances and the EV charging.
6. The PV plant is shared by the community of prosumers. The PV panels operate in a horizontal position and at the point of maximum power.
7. The effect of the presence of clouds on the yield of the PV plant is not considered.

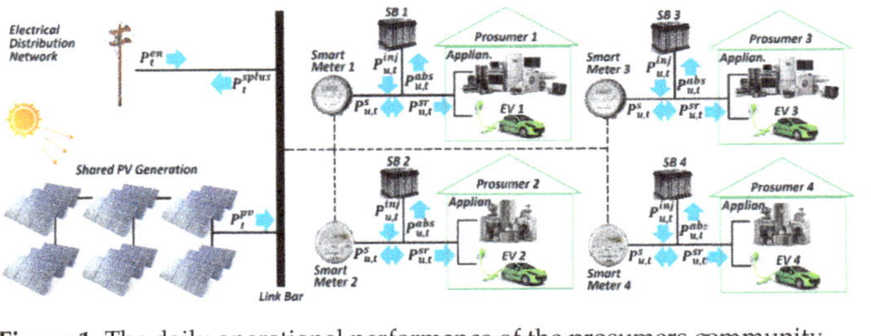


**Figure 1.** The daily operational performance of the prosumers community.

**Table 1.** Technical Data related to Home Appliances with $\beta_a = 1$.

| $a$ | **Appliances** | $P_a$ | $\tau_a$ | $\underline{Q}_a$ | $\underline{\tau}_a$ | $Q_a$ | $\beta_a^i$ |
|---|---|---|---|---|---|---|---|
| 1 | Air Conditioner | 4.00 | 2 | 2 | 0.25 | - | 1 |
| 2 | Freezer | 0.40 | 10 | 10 | 0.50 | - | 0 |
| 3 | Clothes Dryer | 3.50 | 1 | 1 | 0.50 | - | 1 |
| 4 | Computer | 0.25 | 2 | 2 | 0.50 | - | 0 |
| 5 | Incand. light | 0.10 | 5 | 5 | 0.25 | - | 0 |
| 6 | TV | 0.09 | 5 | 5 | 0.50 | - | 0 |
| 7 | Electric Iron | 1.00 | 1 | 1 | 0.25 | - | 1 |
| 8 | Fan | 0.10 | 4 | 4 | 0.50 | - | 0 |
| 9 | DVD Player | 0.025 | 2 | 2 | 0.25 | - | 0 |
| 10 | Stereo | 0.020 | 2 | 2 | 0.25 | - | 0 |

**Table 2.** Technical Data related to Home Appliances with $\beta_a = 0$.

| $a$ | **Appliances** | $P_a$ | $\tau_a$ | $\underline{Q}_a$ | $\underline{\tau}_a$ | $Q_a$ | $\beta_a^i$ |
|---|---|---|---|---|---|---|---|
| 11 | Electric Faucet | 3.50 | 0.50 | 1 | - | 1 | 1 |
| 12 | Dishwasher | 1.50 | 0.75 | 1 | - | 1 | 1 |
| 13 | Coffee Maker | 1.00 | 0.50 | 1 | - | 1 | 0 |
| 14 | Resistance Oven | 1.50 | 0.50 | 1 | - | 1 | 0 |
| 15 | Electric Shower | 3.50 | 0.15 | 1 | - | 1 | 1 |
| 16 | Microwave | 1.30 | 0.33 | 1 | - | 1 | 0 |
| 17 | Washing Machine | 1.50 | 0.50 | 1 | - | 1 | 0 |
| 18 | Vacuum Cleaner | 1.00 | 0.33 | 1 | - | 1 | 0 |
| 19 | Hair Dryer | 0.70 | 0.50 | 1 | - | 1 | 0 |
| 20 | Toaster | 0.80 | 0.16 | 1 | - | 1 | 0 |

**Table 3.** Technical Data related to EVs ($\beta_a = -1$).

| $a$ | **Appliances** | $P_a$ | $C_u^{ev}$ | $\tau_u^{ev}$ | $\underline{Q}_u^{ev}$ | $\beta_a^i$ |
|---|---|---|---|---|---|---|
| 21 | EV 1, 2, 3 | 4.00 | 20.0 | 0.5 | 5.00 | 1 |

## 2.2. Shared PV Plant and Prosumers Community Operation

In the shared PV plant, the solar radiation, $\mathcal{A}^{pv} \times \mathcal{G}_t$, is converted to electrical power in DC, $P_t^{pv}$, using (1).

$$P_t^{pv} = \zeta^p \times \zeta^{ti} \times \zeta^{dd} \times \zeta^{ml} \times \zeta^{cl} \times \mathcal{A}^{pv} \times \mathcal{G}_t, \forall t \in T \quad (1)$$

The power conversion is performed considering the effect of the tolerance levels on the panel production, temperature, dirt and dust, power losses in the cabling, and power losses on the DC and AC sides represented by the coefficients $\zeta^p$, $\zeta^{ti}$, $\zeta^{dd}$, $\zeta^{ml}$, and $\zeta^{cl}$ respectively [43,44]. Note that the PV power produced by the shared plant depends on the levels of solar radiation throughout the day, $\mathcal{G}_t$ [45,46]. It is worth mentioning that in a more realistic scenario, the PV power produced by the shared plant can be reduced to more than 40% of its nominal power (depending on the number of connected arrays) due to cloud displacement during the day [47–49]. In this sense, in our proposed model, once introduced as a feature in the smart meter, it must evaluate the stochastic behavior of solar radiation (e.g., due to the presence of clouds, etc.) and determine, for each period $t$, an average of solar radiation values (based on the weather data). Therefore, each radiation value in period t can be assigned to $\mathcal{G}_t$ in (1) and thus our proposal can be applied in real systems.

The daily operation of the prosumers' community in the SG environment is depicted in Figure 1. In this scheme, energy storage devices (such as SB 1, SB 2, SB 3, and SB 4), which support the energy consumption of each prosumer $u$, consider the power flows $P_{u,t}^{abs}$

and $P^{inj}_{u,t}$ in each period $t$. Note that both powers are the components of $P^{s}_{u,t}$ depending on the state of each SB (recharge/discharge) in period $t$. In addition, the energy stored in the SB, $C^{sb}_{u}$, during the day considers the $\eta^{abs}_{u}$ and $\eta^{inj}_{u}$ rates. Also, note that the power $P^{s}_{u,t}$ has two directions (i.e., from the grid to the smart home or from the smart home to the power grid) depending on the $C^{sb}_{u}$ in the SB, the power generation $P^{pv}_{t}$ of the PV plant, and the electricity price, $\zeta_t$. Household appliances $a$, including the EV of each prosumer $u$ are supplied through the power $P^{sr}_{t}$ throughout the day. Each smart home has a habitual consumption profile represented by $\mathcal{H}^{cp}_{u,a,t}$, which is optimized through the MILP model, obtaining $\mathcal{O}^{cp}_{u,a,t}$.

### 2.3. Habitual Consumption Profile and Hourly Preferences

Hourly preferences and the habitual usage patterns of each appliance are two key aspects causing changes in the consumption profile of customers. The habitual consumption profile considers the usage habits of each appliance $a$, existing in a smart home, according to the needs of the prosumer $u$ throughout the day. In each smart home, a number of appliances equal to twenty-one are being considered, including the EV. Tables 1–3 present the technical characteristics of each appliance $a$ [50,51]. Appliances $a$ with working time greater than 1 h are shown in Table 1. In Table 2, all appliances $a$ with working hours less than 1 h are listed. Finally, Table 3 reports the technical data for EVs. Due to the variable nature of customers' needs during the day and the day-to-day, uncertainties in the usage of household appliances $a$ in given periods $t$ need to be considered for having a more practical model. Algorithm 1, which is based on the Monte Carlo method [52], is used to simulate these uncertainties.

Algorithm 1 starts considering the date related to $P_a$, $\mathfrak{C}_{a,t}$, $\underline{Q}_a$, $Q_a$, $Q^{ev}_{u}$, and $\Delta t$, which are the average power of each appliance $a$, the probability of using each appliance $a$ in the period $t$ [53], the minimum number of times that a given appliance $a$ with $\beta_a = 1$ is turned on, the number of times that a given appliance $a$ with $\beta_a = 0$ is turned on, the number of times that EV with $\beta_a = -1$ is connected for battery charging, and the duration of each period $t$. The values to be adopted by indexes $u$, $a$, and $t$ are established through intervals $1 \ldots |U|$, $1 \ldots |A|$, and $1 \ldots |T|$, respectively. Then, the values of $\hat{\mathfrak{C}}_{a,t}$, $X^{hp}_{u,a,t}$, and $\Psi_{u,a}$ are initialized; $\hat{\mathfrak{C}}_{a,t}$ and $X^{hp}_{u,a,t}$ to zero, and the value of $\Psi_{u,a}$ depends on $\beta_a$. Note that for values of $\beta_a$ such as 1, 0, and $-1$, the parameter $\Psi_{u,a}$ assumes the values $\underline{Q}_a$, $Q_a$, and $Q^{ev}_{u}$, respectively. After that, an iterative process related to each appliance $a$ is performed. For each iteration, the value of the accumulator $k$ is set to zero. Additionally, for each iteration $a$, another iterative process for each period $t$ is done. Within this process, $\mathfrak{C}_{a,t}$ is added to the current value of $k$, thus obtaining a new value for $k$, which in turn is assigned to $\hat{\mathfrak{C}}_{a,t}$. Both iterative processes are completed once the indexes $a$ and $t$ reach the values of $|A|$ and $|T|$, respectively. Hereafter, an iterative process related to each prosumer $u$ is performed. Next, another iterative process for each appliance $a$ is also performed. Within this last iterative process, an infinite loop is executed. For each iteration, a random number, $\rho$, between zero and one hundred is generated. Then, within the iterative process related to periods $t$, the random value $\rho$ is evaluated through the condition $\hat{\mathfrak{C}}_{a,t-1} \leq \rho \leq \hat{\mathfrak{C}}_{a,t}$. If this condition is met, then a value of one is assigned to $X^{hp}_{u,a,t}$, otherwise, $X^{hp}_{u,a,t}$ remains zero. Once all values of $t$ are completed, a new condition is evaluated. Thus, if the condition $\sum_{t=1}^{|T|} X^{hp}_{u,a,t} = \Psi_{u,a}$ is verified, then, another iterative process for each period $t$ is performed, and, in each iteration $t$, the value of $\mathcal{H}^{cp}_{u,a,t}$ is calculated as the product $P_a \times X^{hp}_{u,a,t} \times \Delta t$. Still in the same condition with $\sum_{t=1}^{|T|} X^{hp}_{u,a,t} = \Psi_{u,a}$ being verified, after index $t$ has completed the total number of iterations, $|T|$, the execution of the algorithm goes back to step 12, and the infinite loop stops. When the previous condition is not met, the algorithm returns to step 18, and then the infinite loop also ends. The algorithm ends when $u$ and $a$ have completed all their values.

**Algorithm 1.** Simulation of uncertainties in household appliances usage.

1. Set parameters: $P_a$, $\mathfrak{C}_{a,t}$, $\underline{Q}_a$, $Q_a$, $Q_u^{ev}$, $\Delta t$, $u \in 1...|U|$, $a \in 1...|A|$, $t \in 1...|T|$
2. Values to initialize: $\widehat{\mathfrak{C}}_{a,t} \leftarrow 0$, $X_{u,a,t}^{hp} \leftarrow 0$, $u \in 1...|U|$, $a \in 1...|A|$, $t \in 1...|T|$
3. $\Psi_{u,a} \leftarrow \begin{cases} \underline{Q}_a;\ if\ \beta_a = 1 \\ Q_a;\ if\ \beta_a = 0 \\ Q_u^{ev};\ if\ \beta_a = -1 \end{cases}$, $u \in 1...|U|$, $a \in 1...|A|$
4. **For** $a = 1$ to $|A|$ **do**
5.     set $k \leftarrow 0$;
6.     **For** $t = 1$ to $|T|$ **do**
7.         set $k \leftarrow k + \mathfrak{C}_{a,t}$;
8.         obtain $\widehat{\mathfrak{C}}_{a,t} \leftarrow k$;
9.     **End for**
10. **End for**
11. **For** $u = 1$ to $|U|$ **do**
12.     **For** $a = 1$ to $|A|$ **do**
13.         **While** (true) **do**
14.             $\rho \leftarrow$ rand (0, 100);
15.             **For** $t = 1$ to $|T|$ **do**
16.                 **if** $\widehat{\mathfrak{C}}_{a,t-1} \le \rho \le \widehat{\mathfrak{C}}_{a,t}$ **then**
17.                     Let $X_{u,a,t}^{hp} \leftarrow 1$;
18.                 **End if**
19.             **End for**
20.             **if** $\sum_{t=1}^{|T|} X_{u,a,t}^{hp} = \Psi_{u,a}$ **then**
21.                 **For** $t = 1$ to $|T|$ **do**
22.                     Calculate $\mathcal{H}_{u,a,t}^{cp} \leftarrow P_a \times X_{u,a,t}^{hp} \times \Delta t$;
23.                 **End for**
24.                 Break; back to step 12;
25.             **Else**
26.                 Break; back to step 18;
27.             **End if**
28.         **End while**
29.     **End for**
30. **End for**

As a result of the execution of this Algorithm 1, the values of the habitual consumption profile, $\mathcal{H}_{u,a,t}^{cp}$, are obtained for each prosumer $u$, for each appliance $a$, and in each period $t$. Moreover, the $\mathcal{H}_{u,a,t}^{cp}$ values are part of the MILP model input data.

The hourly preferences $\theta_{u,a,t}^{p}$ of household appliances $a$ are understood as flexibility, i.e., postponing or anticipating the periods $t$ when a given appliance $a$ can be turned on for consumption without compromising the comfort or lifestyle of each residential prosumer $u$. In this work, these periods $t$ of flexibility in the household appliances usage are represented through the binary parameter $\theta_{u,a,t}^{p}$. Thus, when $\theta_{u,a,t}^{p}$ adopts the value of 1, then the corresponding period $t$ is part of the hourly preferences related to the appliance $a$, otherwise, periods $t$ with $\theta_{u,a,t}^{p} = 0$ do not correspond to the hourly preferences but may have the habitual consumption of electricity, $\mathcal{H}_{u,a,t}^{cp}$. For example, Figure 2 shows the $\theta_{u,a,t}^{p} = 1$ values for air conditioning as well as the EV during the day. In the case of air conditioning, prosumers can switch on their appliance at certain times, including peak hours. For EVs, the night and dawn periods are considered. For all prosumers $u$, the $\theta_{u,a,t}^{p}$ values for each appliance $a$ in each period $t$ are assumed to be [54], taking into account the criteria of [12,50,51,55].

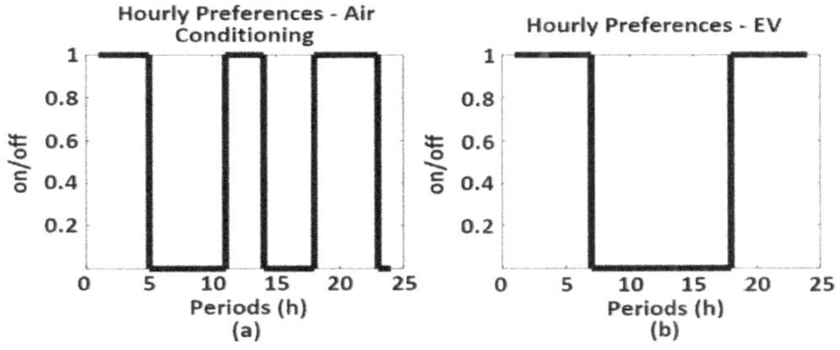


**Figure 2.** Hourly preferences related to air conditioning and the EV.

## 3. Mathematical Model

The initial formulation of the problem is represented by a mixed-integer nonlinear programming (MINLP) model. Nonlinear terms are presented in the objective function as well as in the operational constraints. These constraints are related to household appliances, including each EV presented in homes, as well as the power balance of the electrical system and the energy stored through the SB.

### 3.1. Objective Function

The objective function (2) is composed of two functions, $\psi_1$ and $\psi_2$. In function $\psi_1$, the terms starting with the weighting coefficients $\varpi^{(1)}$, $\varpi^{(2)}$, and $\varpi^{(3)}$ are respectively related to the reduction in costs per energy consumption, coincident usage of household appliances with higher average power, and electricity purchase from the utility company. Therefore, $\psi_1$ aims to minimize the key elements that have a direct effect on prosumer's electricity bill. On the other hand, the function $\psi_2$ allows reshaping of the optimal consumption profile by reducing the difference between $P_t^{at}$ and $P^{av}$, which implicitly maximizes the value of LF.

$$\text{Minimize } f = \psi_1 + \psi_2 \tag{2}$$

where,

$$\psi_1 = \varpi^{(1)} \times \left[ \sum_{\forall u \in U} \sum_{\forall a \in A} \sum_{\forall t \in T} \zeta_t \times O_{u,a,t}^{cp} \right] + \varpi^{(2)} \times \left[ \sum_{\forall u \in U} \sum_{\forall t \in T} \theta_{u,t}^{i} \right] + \varpi^{(3)} \times \left[ \sum_{\forall t \in T} P_t^{en} \right]$$

$$\psi_2 = \varpi^{(4)} \times \left[ \sum_{\forall t \in T} \left( P_t^{at} - P^{av} \right)^2 \right]$$

### 3.2. Constraints

#### 3.2.1. Home Appliances Constraints

The constraints described below are related to the operating regime of household appliances as well as the charging of EV batteries and are based on the research [40]. Equations (3) and (4), through variable $X_{u,a,t}^{op}$, determine for each period $t$, the on/off state of each appliance $a$ belonging to the prosumer $u$. In (3), the off state of each appliance $a$ is established only for hourly preferences with $\theta_{u,a,t}^{p} = 0$. For each prosumer $u$ with hourly preference values $\theta_{u,a,t}^{p} = 1$, (4) indicates whether the appliance $a$ can be turned on or off in a given period $t$. Equation (5) is related to household appliances with higher average power, $\beta_a^i = 1$. Thus, for each prosumer $u$, the value of $\theta_{u,t}^{i}$ calculates the number of appliances $a$ in each period $t$ with $\beta_a^i = 1$ that can be turned on.

$$X_{u,a,t}^{op} = 0, \forall u \in U, \forall a \in A, \forall t \in T \mid \theta_{u,c,t}^{p} = 0 \wedge t \geq 1 \tag{3}$$

$$X_{u,a,t}^{op} \leq 1, \forall u \in U, \forall a \in A, \forall t \in T \mid \theta_{u,a,t}^{p} = 1 \wedge t \geq 1 \tag{4}$$

$$\theta^{i}_{u,t} = \sum_{\forall a \in A} \left( X^{op}_{u,a,t} \times \theta^{p}_{u,a,t} \times \beta^{i}_{a} \right), \forall u \in U, \forall t \in T \mid t \geq 1 \quad (5)$$

The operating regime of household appliances with working hours greater than or equal to 1 h, $\beta_a = 1$, is modeled by (6)–(9). The number of times this type of appliance is turned on during the day is established by the limits $\underline{Q}_a$ and $\overline{Q}_a$ in (6), while the time of usage of the appliance $a$ within each period $t$ is obtained by (7). The optimal consumption profile, $\mathcal{O}^{cp}_{u,a,t}$, in each period $t$ in which appliances $a$ with $\beta_a = 1$ are turned on is calculated using (8). In (9), for each prosumer $u$, the total time of usage of the appliance $a$ during the day, $\tau_a$, is guaranteed through the product of $\theta^{p}_{u,a,t}$, $X^{op}_{u,a,t}$ and $\tau^{us}_{u,a,t}$.

$$\underline{Q}_a \leq \sum_{\forall t \in T \mid t \geq 1} X^{op}_{u,a,t} \times \theta^{p}_{u,a,t} \leq \overline{Q}_a, \forall u \in U, \forall a \in A \mid \beta_a = 1 \quad (6)$$

$$\underline{\tau}_a \leq \tau^{us}_{u,a,t} \leq \overline{\tau}_a, \forall u \in U, \forall a \in A, \forall t \in T \mid \beta_a = 1 \wedge t \geq 1 \quad (7)$$

$$\mathcal{O}^{cp}_{u,a,t} = P_a \times \theta^{p}_{u,a,t} \times X^{op}_{u,a,t} \times \tau^{us}_{u,a,t}, \forall u \in U, \forall a \in A, \forall t \in T \mid \beta_a = 1 \wedge t \geq 1 \quad (8)$$

$$\sum_{\forall t \in T \mid t \geq 1} X^{op}_{u,a,t} \times \theta^{p}_{u,a,t} \times \tau^{us}_{u,a,t} = \tau_a, \forall u \in U, \forall a \in A \mid \beta_a = 1 \quad (9)$$

Constraints (10)–(11) are related to appliances $a$ with working hours less than 1 h, i.e. type of appliance with $\beta_a = 0$. In (10), the number of times that each appliance $a$ is utilized throughout the day is established, while the optimal consumption, $\mathcal{O}^{cp}_{u,a,t}$, of each appliance $a$ related to the prosumer $u$ is obtained by (11).

$$\sum_{\forall t \in T \mid t \geq 1} \theta^{p}_{u,a,t} \times X^{op}_{u,a,t} = Q_a, \forall u \in U, \forall a \in A \mid \beta_a = 0 \quad (10)$$

$$\mathcal{O}^{cp}_{u,a,t} = P_a \times \theta^{p}_{u,a,t} \times X^{op}_{u,a,t} \times \tau_a, \forall u \in U, \forall a \in A, \forall t \in T \mid \beta_a = 0 \wedge t \geq 1 \quad (11)$$

Through (12)–(18), the operating regime of EV (the type of appliance $a$ with $\beta_a = -1$ related to the prosumer $u$ is represented. For each prosumer $u$, the energy stored in the EV battery in each period $t$ is determined in (12). In (13), the variation interval related to the charging time of the EV battery is established considering the limits $[\underline{\tau}^{ev}_u, \overline{\tau}^{ev}_u]$, while (14) ensures the number of times that the EV battery can be charged within the range $\left[\underline{Q}^{ev}_u, \overline{Q}^{ev}_u\right]$. The optimal profile, $\mathcal{O}^{cp}_{u,a,t}$, of the EV ($\beta_a = -1$) related to prosumer $u$ is obtained by (15) as the total energy stored in the EV battery during the day is calculated using (16). This total energy, $\varepsilon^{ev}_u$, must be equal to a percentage $\lambda_1$ of the battery capacity, $\mathcal{C}^{ev}_u$, of the EV, as (17).

$$SoC^{ev}_{u,t} = P^{ev} \times X^{op}_{u,a,t} \times \tau^{ev}_{u,t}, \forall u \in U, \forall a \in A, \forall t \in T \mid \beta_a = -1 \wedge t \geq 1 \quad (12)$$

$$\underline{\tau}^{ev}_u \leq \tau^{ev}_{u,t} \leq \overline{\tau}^{ev}_u, \forall u \in U, \forall t \in T \mid t \geq 1 \quad (13)$$

$$\underline{Q}^{ev}_u \leq \sum_{\forall t \in T \mid t \geq 1} X^{op}_{u,a,t} \leq \overline{Q}^{ev}_u, \forall u \in U, \forall a \in A \mid \beta_a = -1 \quad (14)$$

$$\mathcal{O}^{cp}_{u,a,t} = SoC^{ev}_{u,t}, \forall u \in U, \forall a \in A, \forall t \in T \mid \beta_a = -1 \wedge t \geq 1 \quad (15)$$

$$\varepsilon^{ev}_u = SoC_u + \sum_{\forall t \in T \mid t \geq 1} SoC^{ev}_{u,t}, \forall u \in U \quad (16)$$

$$\varepsilon^{ev}_u = \lambda^1 \times \mathcal{C}^{ev}_u, \forall u \in U \quad (17)$$

$$\sum_{\forall a \in A} \mathcal{O}^{cp}_{u,a,t} \leq Max_{\forall t' \in T} \left[ \sum_{\forall a \in A} \mathcal{H}^{cp}_{u,a,t'} \right], \forall u \in U, \forall t \in T \mid t \geq 1 \quad (18)$$

After calculating $\mathcal{O}^{cp}_{u,a,t}$ of each appliance $a$ related to the prosumer $u$, (18) ensures that the total consumption in each period $t$ of the optimal profile $\mathcal{O}^{cp}_{u,a,t}$ not exceed the peak consumption of the habitual profile $\mathcal{H}^{cp}_{u,a,t}$.

3.2.2. Power Balance Constraints

Constraints (19)–(24) are related to the power flow between the PV plant, the electricity company, the smart home of each prosumer (including EV), as well as the SB, as depicted in Figure 1. The power balance is guaranteed in (19). Note that this equation considers the surplus of energy injected into the electrical network, either by the PV plant or by the SBs in each period $t$. The non-negativity of the power coming from the electric grid, $P_t^{en}$, is guaranteed in (20). In (21), the power $P_{u,t}^{s}$ is calculated for each period $t$ as the sum of power supplied to the smart home of the prosumer $u$, $P_{u,t}^{sr}$, and the $P_{u,t}^{inj}$ / $P_{u,t}^{abs}$ powers, related to the SB. The $P_{u,t}^{sr}$ values for the prosumer $u$ in each period $t$ are computed in (22). The total power required by all smart homes, $P_t^{at}$, in each period $t$ is obtained using (23). Finally, (24) determines the average value of $P_t^{at}$ that is represented by $P^{av}$. It is worth noting that both (23) and (24) are linked to the improvement of the LF, as can be seen in (2).

$$P_t^{pv} + \left(P_t^{en} - P_t^{splus}\right) = \sum_{\forall u \in U} P_{u,t}^{s},\ \forall t \in T \mid t \geq 1 \tag{19}$$

$$0 \leq P_t^{en},\ \forall t \in T \mid t \geq 1 \tag{20}$$

$$P_{u,t}^{s} = P_{u,t}^{sr} + \left(\eta_u^{abs} \times P_{u,t}^{abs} - \eta_u^{inj} \times P_{u,t}^{inj}\right), \forall u \in U,\ \forall t \in T \mid t \geq 1 \tag{21}$$

$$P_{u,t}^{sr} = \left(\frac{1}{\Delta t}\right) \times \sum_{\forall a \in A} \mathcal{O}_{u,a,t}^{cp}, \forall u \in U,\ \forall t \in T \mid t \geq 1 \tag{22}$$

$$P_t^{at} = \sum_{\forall u \in U} P_{u,t}^{sr},\ \forall t \in T \mid t \geq 1 \tag{23}$$

$$P^{av} = \left(\frac{1}{|T|}\right) \times \sum_{\forall t \in T \mid t \geq 1} P_t^{at} \tag{24}$$

3.2.3. Energy Storage Constraints

Constraints (25)–(31) represent the operation of the SBs related to each prosumer $u$. The injected power $P_{u,t}^{inj}$ and the absorbed power $P_{u,t}^{abs}$ by the SB are limited by (25) and (26), respectively. The injection/absorption status of the SB is established by (27). Note that when the SB is injecting power ($\mu_{u,t}^{inj}$ = 1), it cannot absorb power ($\mu_{u,t}^{abs}$ = 1), and vice versa. In (28), the initial state of charge $SoC_{u,t}^{sb}$ ($t$ = 0) of the batteries is established as a percentage $\lambda_u^2$ of the storage capacity, of each SB related to the prosumer $u$, $C_u^{sb}$. It is worth mentioning that the period $t$ = 0 is related to the last period of the previous day. The state of charge of the SB for the rest of the periods $t$ is obtained by (29). Constraint (30) guarantees that at the end of the day $t$ = $|T|$ the $SoC_{u,t}^{sb}$ of the SB is equal to its storage capacity, $C_u^{sb}$, while (31) guarantees that the SoC of SB throughout the day does not exceed the $C_u^{sb}$ capacity.

$$\mu_{u,t}^{inj} \times \underline{P}_u^{inj} \leq P_{u,t}^{inj} \leq \mu_{u,t}^{inj} \times \overline{P}_u^{inj}, \forall u \in U,\ \forall t \in T \mid t \geq 1 \tag{25}$$

$$\mu_{u,t}^{abs} \times \underline{P}_u^{abs} \leq P_{u,t}^{abs} \leq \mu_{u,t}^{abs} \times \overline{P}_u^{abs}, \forall u \in U,\ \forall t \in T \mid t \geq 1 \tag{26}$$

$$\mu_{u,t}^{inj} + \mu_{u,t}^{abs} = 1, \forall u \in U,\ \forall t \in T \mid t \geq 1 \tag{27}$$

$$SoC_{u,t}^{sb} = \lambda_u^2 \times C_u^{sb}, \forall u \in U,\ \forall t \in T \mid t = 0 \tag{28}$$

$$SoC_{u,t}^{sb} = SoC_{u,t-1}^{sb} + \left(\eta_u^{abs} \times P_{u,t}^{abs} \times \Delta t - \eta_u^{inj} \times P_{u,t}^{inj} \times \Delta t\right), \forall u \in U,\ \forall t \in T \mid t \geq 1 \tag{29}$$

$$SoC_{u,t}^{sb} = C_u^{sb}, \forall u \in U,\ \forall t \in T \mid t = |T| \tag{30}$$

$$0 \leq SoC_{u,t}^{sb} \leq C_u^{sb}, \forall u \in U,\ \forall t \in T \mid t \geq 1 \tag{31}$$

### 3.3. Linearization

As the original model is a mixed-integer nonlinear one, commercial solvers face difficulties in finding the global solution. To address this issue, linearization techniques are applied to recast the nonlinear terms in the objective function $\psi_2$, as well as in constraints (8), (9), and (12).

The Big-M method [56] is applied to linearize (8) and (9). Thus, these equations are replaced by (32)–(35).

$$\mathcal{O}^{cp}_{u,a,t} = P_a \times \theta^{p}_{u,a,t} \times \Delta\tau^{us}_{u,a,t},\ \forall u \in U,\ \forall a \in A,\ \forall t \in T \mid \beta_a = 1 \wedge t \geq 1 \tag{32}$$

$$\sum_{\forall t \in T \mid\ t \geq 1} \theta^{p}_{u,a,t} \times \Delta\tau^{us}_{u,a,t} = \tau_a, \forall u \in U,\ \forall a \in A \mid \beta_a = 1 \tag{33}$$

$$0 \leq -\Delta\tau^{us}_{u,a,t} + \tau^{us}_{u,a,t}\ \leq\ M \times \left(1 - X^{op}_{u,a,t}\right),\ \forall u \in U,\ \forall a \in A,\ \forall t \in T \mid t \geq 1 \tag{34}$$

$$0 \leq \Delta\tau^{us}_{u,a,t} \leq\ M \times X^{op}_{u,a,t},\ \forall u \in U,\ \forall a \in A,\ \forall t \in T \mid t \geq 1 \tag{35}$$

Likewise, (36)–(38) present the linear representation of constraint (12).

$$SoC^{ev}_{u,t} = P^{ev} \times \Delta\tau^{ev}_{u,t}, \forall u \in U,\ \forall a \in A,\ \forall t \in T \mid \beta_a = -1 \wedge t \geq 1 \tag{36}$$

$$0 \leq -\Delta\tau^{ev}_{u,t} + \tau^{ev}_{u,t}\ \leq\ M \times \left(1 - X^{op}_{u,a,t}\right),\ \forall u \in U,\ \forall a \in A,\ \forall t \in T \mid t \geq 1 \tag{37}$$

$$0 \leq \Delta\tau^{ev}_{u,t} \leq\ M \times X^{op}_{u,a,t},\ \forall u \in U,\ \forall a \in A,\ \forall t \in T \mid t \geq 1 \tag{38}$$

To linearize the function $\psi_2$, a discretization process related to the square of the difference between $P^{at}_t$ and $P^{av}$ is performed [57,58]. As a result, function $\psi_2$ is presented by $\psi'_2$, while (39)–(42) are also considered.

$$\begin{gathered} \psi'_2 = \varpi^{(4)} \times \left[\sum_{\forall t \in T}\ \sum_{\forall y \in 1Y} \Pi_{t,\ y} \times \Delta\Lambda_{t,\ y}\right] \\ \Lambda_t = P^{at}_t - P^{av},\ \forall t \in T \mid t \geq 1 \end{gathered} \tag{39}$$

$$\Lambda^{+}_t - \Lambda^{-}_t = \Lambda_t,\ \forall t \in T \mid t \geq 1 \tag{40}$$

$$\Lambda^{+}_t + \Lambda^{-}_t = \sum_{y=1}^{Y} \Delta\Lambda_{t,\ y},\ \forall t \in T \mid t \geq 1 \tag{41}$$

$$0 \leq \Delta\Lambda_{t,\ y} \leq \overline{\Delta}_t,\ \forall t \in T,\ y \in 1 \ldots Y \mid t \geq 1 \tag{42}$$

### 3.4. Linearized Model

The obtained mixed-integer linear programming (MILP) model is presented as follows.
Minimize $f = \psi_1 + \psi'_2$
s.t.
Constraints: (3)–(7), (10), (11), (13)–(18), (19)–(24), (25)–(31), (32)–(35), (36)–(38), (39)–(42).

## 4. Results and Discussion

This section presents the simulation results and the corresponding discussions to validate the proposed framework related to intelligent energy management in a prosumers community, considering the LF improvement. This model was written in the algebraic language AMPL and solved using the solver CPLEX on an Intel(R) Core(TM) i5 CPU M480@2.67GHz personal computer with a 4.00 GB RAM and a 64-bit Windows 7 operating system.

### 4.1. Basic Data

The proposed model is evaluated using the information provided in Tables 1–3 related to the technical data of household appliances $a$, as well as the hourly preferences $\theta^{p}_{u,a,t}$ whose periods $t$ are specified in [54]. By executing Algorithm 1, the values of the habitual consumption profile, $\mathcal{H}^{cp}_{u,a,t}$, are simulated for each prosumer $u$, each appliance $a$, and for all periods $t$. Then, the proposed MILP model is executed based on the information mentioned above to obtain the optimal consumption profile, $\mathcal{O}^{cp}_{u,a,t}$, result in efficient scheduling of household appliances $a$, as well as the optimal performance of energy generation and storage technologies, i.e., shared PV plant and SBs. The efficient scheduling of household appliances $a$ is driven by a scheme tariff $\zeta_t$ [59] that reflects the price of energy in a given period $t$, see Table 4. The EV belonging to each prosumer $u$ starts the day with a state of charge equal to zero and must be charged with a rate of $P^{ev}$ over 4.00 kW until it is fully charged ($\lambda^1 = 1$ and $C^{ev}_u = 20$ kWh) [56]. The operation of the PV plant and storage batteries is derived from [60–63]. In the shared PV plant, the power is produced according to (1). Here, the $\mathcal{G}_t$ values during the day are obtained considering the maximum local solar radiation $\hat{\mathcal{G}}$ equal to 0.22 kW/m$^2$ [64], as well as the per-unit values of standard solar radiation $\gamma_t$ [65]. This PV power production also considers the loss factors that are associated with energy conversion, such as $\zeta^{p} = 0.95$, $\zeta^{ti} = 0.89$, $\zeta^{dd} = 0.93$, $\zeta^{ml} = 0.95$, and $\zeta^{cl} = 0.90$. Moreover, the effective PV area of 116.64 m$^2$ considers an arrangement of 60 panels of 330 kW [66]. For SBs, the energy storage capacity, $C^{sb}_u$, of prosumers 1, 2, and 3 are 12.0 kWh, 10.0 kWh, and 15.0 kWh, respectively. The SBs start their operation with $\lambda^2_u$ = 35%, 45%, and 25% of the capacity $C^{sb}_u$, for each prosumer $u$ [62]. Limits $[\underline{P}^{inj}_u, \overline{P}^{inj}_u]$ of the power injected by each SB are [1.00, 3.00] for prosumer 1, [1.50, 3.50] for prosumer 2, and [0.50, 2.50] for prosumer 3. Also, limits $\left[\underline{P}^{abs}_u, \overline{P}^{abs}_u\right]$ of the power absorbed by SBs are assumed to be [1.00, 3.50] for prosumer 1, [0.50, 4.00] for prosumer 2, and [0.50, 3.00] for prosumer 3. In addition, the recharging and discharging of SB is performed with efficiencies $\eta^{abs}_u$ and $\eta^{inj}_u$ of 0.98 and 0.99, respectively [63]. The weighting factors $\varpi^{(1)}$, $\varpi^{(2)}$, $\varpi^{(3)}$, and $\varpi^{(4)}$ related to the functions $\psi_1$ and $\psi_2$ adopt the values 1.0, 10.0, 100.0, and 100.0, respectively. Note that the weights $\varpi^{(3)}$ and $\varpi^{(4)}$ have a higher value and are related to the power supplied by the energy company and the term related to the improvement of the LF. The constants $M$, and the number of discretization blocks $|Y|$ are set to 1000, and 5, respectively. Also, the value of $\overline{\Delta}_t$ is calculated as $10/|Y|$ in each period $t$ [57].

**Table 4.** Hourly values related to the price of energy, standard and local solar radiation.

| t | Periods | $\zeta_t$ (\$/kWh) | $\gamma_t$ (p.u.) | $\mathcal{G}_t$ (kW)/m$^2$ | t | Periods | $\zeta_t$ (\$/kWh) | $\gamma_t$ (p.u.) | $\mathcal{G}_t$ (kW)/m$^2$ |
|---|---|---|---|---|---|---|---|---|---|
| 1 | 01:00–02:00 h | 0.22419 | 0.00 | 0.00 | 13 | 13:00–14:00 h | 0.22419 | 1.00 | 0.220 |
| 2 | 02:00–03:00 h | 0.22419 | 0.00 | 0.00 | 14 | 14:00–15:00 h | 0.22419 | 0.95 | 0.209 |
| 3 | 03:00–04:00 h | 0.22419 | 0.00 | 0.00 | 15 | 15:00–16:00 h | 0.22419 | 0.82 | 0.180 |
| 4 | 04:00–05:00 h | 0.22419 | 0.00 | 0.00 | 16 | 16:00–17:00 h | 0.22419 | 0.53 | 0.117 |
| 5 | 05:00–06:00 h | 0.22419 | 0.00 | 0.00 | 17 | 17:00–18:00 h | 0.22419 | 0.15 | 0.033 |
| 6 | 06:00–07:00 h | 0.22419 | 0.10 | 0.022 | 18 | 18:00–19:00 h | 0.32629 | 0.08 | 0.018 |
| 7 | 07:00–08:00 h | 0.22419 | 0.20 | 0.044 | 19 | 19:00–20:00 h | 0.51792 | 0.00 | 0.00 |
| 8 | 08:00–09:00 h | 0.22419 | 0.50 | 0.11 | 20 | 20:00–21:00 h | 0.51792 | 0.00 | 0.00 |
| 9 | 09:00–10:00 h | 0.22419 | 0.80 | 0.176 | 21 | 21:00–22:00 h | 0.51792 | 0.00 | 0.00 |
| 10 | 10:00–11:00 h | 0.22419 | 0.90 | 0.198 | 22 | 22:00–23:00 h | 0.32629 | 0.00 | 0.00 |
| 11 | 11:00–12:00 h | 0.22419 | 0.95 | 0.209 | 23 | 23:00–24:00 h | 0.22419 | 0.00 | 0.00 |
| 12 | 12:00–13:00 h | 0.22419 | 1.00 | 0.220 | 24 | 24:00–01:00 h | 0.22419 | 0.00 | 0.00 |

### 4.2. Simulation Results

Based on the information presented above, the optimal management of electricity consumption in a prosumer community can be performed without considering (Case 1) or with considering (Case 2) the improvement of the LF. In Case 1, the proposed model is executed without taking into account the $\psi'_2$ function as part of (2). Unlike Case 1, Case 2 considers the function $\psi'_2$ to guarantee the improvement of the LF.

Figure 3 shows the profile of habitual, $\mathcal{H}^{cp}_{u,a,t}$, and optimal consumption for each prosumer (in Figure 3a–c) and for all prosumers (community, in Figure 3d). The $\mathcal{H}^{cp}_{u,a,t}$ profile obtained through Algorithm 1 is shown by the red line. The $\mathcal{O}^{cp}_{u,a,t}$ profiles in the blue and black line represent Case 1 and 2, respectively. In Figure 3a, the $\mathcal{H}^{cp}_{u,a,t}$ profile related to prosumer 1 reaches the maximum values of 8.35 kWh and 8.94 kWh in the intermediate period ($t$ = 18) and peak period ($t$ = 20), respectively. In Case 1 the maximum peak is reduced to 6.49 kWh and shifted to period $t$ = 11. Although this consumption profile presents a smooth distribution of consumption throughout the day the maximum consumption peak appears during off-peak hours. Also, other peaks of consumption of 4 kWh are presented in the respective periods of $t$ = 1 and $t$ = 6 with the lowest electricity price. In Case 2, the distribution of consumption during the day is greater than Case 1 previously described. Note that there is not a single maximum peak, but a distributed consumption with values of 3.63 kWh ($t$ = 1), 3.43 kWh ($t$ = 4), 3.88 kWh ($t$ = 5), 3.96 kWh ($t$ = 11), 3.20 kWh ($t$ = 12), and 3.50 kWh ($t$ = 17). Moreover a slight consumption concentration of 4.50 kWh at $t$ = 23 shows how improving the LF contributes to a wide distribution of consumption without creating high peaks. Figure 3b,c related to the consumption profiles $\mathcal{H}^{cp}_{u,a,t}$ of prosumers 2 and 3 show consumption peaks during peak hours ($t$ = 20 with 8.96 kWh, prosumer 3) and in times close to that ($t$ = 23 with 7.60 kWh, prosumer 2). In the same way, as in Figure 3a, the $\mathcal{O}^{cp}_{u,a,t}$ profile for both prosumers, in Case 1, shows a shift from the maximum consumption peak to periods with cheap energy prices, as $t$ = 13 and $t$ = 12 with 5.50 kWh (prosumer 2) and 5.87 kWh (prosumer 3), respectively. For Case 2, the $\mathcal{O}^{cp}_{u,a,t}$ profile related to prosumer 2 presents a peak consumption of 5.43 kWh at $t$ = 19. Within the same period, prosumers 2 and 3 present zero consumption. Also, note that within $t$ = 20 and $t$ = 21, the only consumption belongs to prosumer 3. At this same time, the consumption of prosumers 1 and 2 are reduced such that it does not increase the cost of electricity consumption by the community as well as to avoid congestion during power supply in peak periods. This fact shows the efficiency of the proposed model to schedule the usage of domestic loads, especially those with a higher average power, to obtain a more homogeneous distribution in the consumption profile without harming the lifestyle of the prosumers' community. This homogeneous distribution due to the efficient scheduling of household appliances can be seen in Figure 3d. In this figure, the habitual profile of the community reaches a peak of 18.84 kWh at $t$ = 20. For the $\mathcal{O}^{cp}_{u,a,t}$ profiles, the maximum consumption peaks represent 81.48% (in Case 1) and 40% (in Case 2) of 18.84 kWh, both in the respective periods of $t$ = 12 and $t$ = 1, 8, 11, 12, and 13. Furthermore, the optimal consumption profile Case 1, the blue line, shows the maximum peak at $t$ = 12; in addition to this peak, other consumption peaks are presented within off-peak hours, creating potential load concentrations, which in turn it can cause congestion during power supply. However, in Case 2, for the $\mathcal{O}^{cp}_{u,a,t}$ profile, the black line, this occurrence of new peaks is mitigated. Note that, in this case there is a consumption during peak hours, i.e., the habitual peak consumption has been reduced but not completely eliminated. In Case 1, during peak hours, consumption was completely shifted to other times that possibly contributed to the formation of the new peaks. Therefore, this demonstrates that it is possible to reduce consumption without compromising the reliability of the EDN by considering the improvement in the efficient usage of energy (i.e., increased LF) in intelligent management.

Figure 4a,b demonstrate the values related to the total costs and the LF of each prosumer (1, 2, and 3), as well as the community (total). In both figures, the lines in red, blue, and black are related to the $\mathcal{H}^{cp}_{u,a,t}$ profile, the $\mathcal{O}^{cp}_{u,a,t}$ profile considering Case 1, and the $\mathcal{O}^{cp}_{u,\,a,\,t}$ profile considering Case 2, respectively. For each of these consumption profiles, the respective total costs result in $44.67 (red square), $30.32 (blue square), and $36.47 (black square). Note that the costs of each prosumer $u$ and the total of prosumers related to the Case 1 are lower than those related to Case 2. Financially, Case 1 is more advantageous only for prosumers. This is because energy consumption in the peak period is completely shifted to periods with lower energy prices. However, this schedule of consumption for off-peak periods can contribute to the appearance of new concentrations of consumption in these periods (see Figure 3d), causing disturbances in the reliability of the electrical network, which can increase the maintenance expenses. To alleviate this concern, Case 2, in addition to being an alternative for reducing the community's electricity bill, also reduces the occurrence of peak consumption during off-peak hours (see Figure 3d) through efficient scheduling of home appliances *a* considering hourly preferences, which is advantageous for the energy company and prosumers.

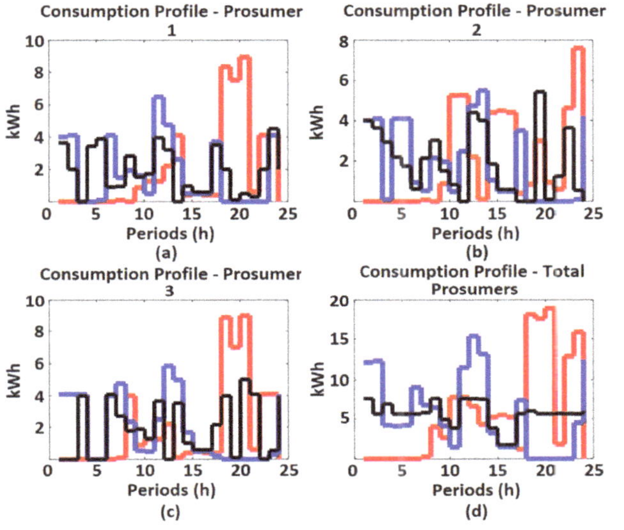


**Figure 3.** Habitual consumption profile and profiles of optimal consumption without and with the improvement of the LF. (**a**) Consumption profiles for prosumer 1; (**b**) Consumption profiles for prosumer 2; (**c**) Consumption profiles for prosumer 3; and (**d**) Consumption profiles for all prosumers.

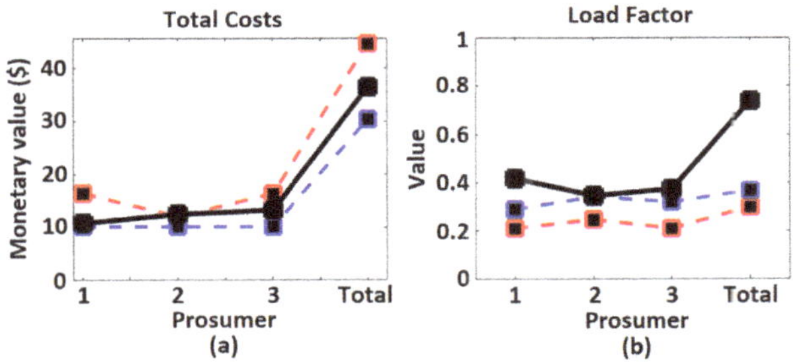


**Figure 4.** Values related to total costs for energy consumption and LF in the prosumers' community. (**a**) Values of total monetary costs; and (**b**) LF values.

Figure 4b shows the values related to the LFs of each prosumer and the community. For profile $\mathcal{H}^{cp}_{u,a,t}$, prosumers 1, 2, and 3 have LF values equal to 0.21, 0.24, and 0.21, while the LF of the community is equal to 0.29. Such values close to zero indicate large concentrations of consumption in given periods of the day, e.g., peak period. Moreover, these low values also indicate that energy is being wasted, i.e., high consumption during peak hours can create stress, especially in the cabling, due to the transformation of electricity into heat that increases energy losses. For Case 1, the LFs have resulted in 0.28, 0.34, and 0.32. Although these LF values are higher than the LF values of the $\mathcal{H}^{cp}_{u,a,t}$ profile, the obtained $\mathcal{O}^{cp}_{u,a,t}$ profiles still present a heterogeneous distribution of consumption during the day. For Case 2, the

LF values of prosumer 1 and 3 have increased while the LF of prosumer 2 has remained the same. Note that the LF values of 0.42, 0.34, and 0.37 related to prosumer 1, 2, and 3 are still close to zero. This fact shows that the proposed strategy has efficiently scheduled the consumption of each customer (Figure 3a–c) in such a way that the consumption profile of all of them (Figure 3d) presents a wide distribution of consumption during the day. Thus, the value of the LF related to the total of prosumer has increased from 0.36 to 0.73 (approximately double).

From Figure 5 onwards, the performance of the PV plant, the energy SBs, and the power flow within the community for Case 1 (blue line) and Case 2 (black line) are analyzed. Figure 5a shows, for each case, the power supplied by the electrical network, $P_t^{en}$, to the community of prosumers in each period $t$. For Case 1, the electric grid supplies the community's energy needs in the interval from $t = 1$ to $t = 6$, and in the period $t = 24$. Note that these periods have cheap energy tariffs. Also, the power grid does not provide power during peak periods, but it is evident that consumption peaks appear at other times, such as at $t = 2$ and $t = 24$, with consumptions equal to 13.72 kWh, and 20.33 kWh, respectively. In Figure 3d, the consumption at $t = 24$ related to this case is less than 20.33 kWh. In this way, part of that consumed energy is provided for each SB presented in a smart home. For Case 2, the power grid supplies energy between periods $t = 2$ to $t = 6$, and $t = 18$ to $t = 24$. Note that in this case, during the peak period, the power supply meets part of the community requirements, specifically 67.83%, while the remainder can be attributed to the energy injected by the SBs. In the same way as in Case 1, a large part of the energy supplied by the electric network in the period $t = 24$ is absorbed by the SBs to guarantee their full charging at the end of the day. Figure 5b shows the values of the surplus power injected into the power grid, $P_t^{splus}$, in both cases. Note that for both cases, most of the energy is injected between periods $t = 9$ to $t = 15$, with full presence of solar radiation, $\mathcal{G}_t$. Moreover, the total amount of surplus energy injected for each case turns out to be 33.32 kWh (Case 1) and 46.73 kWh (Case 2). This fact highlights the effectiveness of Case 2, which ensures the rational usage of energy between the Energy Company and prosumers and adds financial gain to prosumers regarding the sale of PV power produced at the shared plant. Figure 5c shows that the power produced by the shared PV plant is the same for both cases. It is important to note that the surplus power $P_t^{splus}$ in Case 1 and 2 represent 24.20% and 33.94% of the power produced by the PV plant, $P_t^{pv}$, in the day. Although $P_t^{splus}$ is not entirely composed of the plant's PV power, this sold power to the grid contributes to reducing the community's energy bill by approximately 35% when the LF improvement is taken into account.

Figure 6 reports the $P_{u,t}^s$ power values for each prosumer $u$ in each period $t$ of the day. The directions of this power are differentiated by the positive and negative values shown in each case. Positive values indicate that the power $P_{u,t}^s$ measured by the smart meter flows towards the smart home and SB. Otherwise, the power $P_{u,t}^s$ can flow towards the electrical grid or contribute to the energy needs of another smart home in a given period $t$, see Figure 1. For both cases, the number of periods $t$ in the day when the $P_{u,t}^s$ power flows towards the SB and the smart home is much greater than the number of periods $t$ related to the opposite flow. This situation indicates that a minimum part of the energy injected by the SBs is delivered to the electric grid, and the rest of this energy contributes to meet the consumption of household appliances in each smart home. In Case 1, during peak and intermediate periods, the positive values of $P_{u,t}^s$ are minimal. According to Figure 3d, in these periods, energy consumption by household appliances is zero. Thus, the positive values mentioned above represent only the powers absorbed by the SBs. During the same periods, the negative values of $P_{u,t}^s$ represent the power injected by the SBs, which in turn contribute to meeting the energy consumption needs of other smart homes. For example, during peak hours ($t = 20$), it can be seen that there is no presence of solar radiation $\mathcal{G}_t$, and energy price $\zeta_t$ is high. However, the SB related to prosumer 1 absorbs power (equal to 2.94 kW) from the SB related to prosumer 2 (injects 2.44 kW) and 3 (injects 0.50 kW). For Case 2, at $t = 20$, due to the lack of solar radiation, the supply of power to the community is done by the electrical network with a value of $P_t^{en}$ equal to 3.67 kW, as can be seen in Figure 5a. This power is provided for each prosumer $u$ according to the operational status of the SB and the energy consumption of the smart home. For prosumer 1, the demand at $t = 20$ is equal to 0.5 kW (Figure 3a). Being $P_{u,t}^{s}$ positive and equal to 1.48 kW (Figure 6b), the SB absorbs a power $P_{u,t}^{abs}$ of 0.98 kW (Figure 7b). For prosumer 2, the power $P_{u,t}^{s}$ is negative, with a value equal to 3.33 kW (Figure 6d). Furthermore, in the same period, prosumer 2 has a low demand of 0.1 kW (Figure 3b), causing the SB to inject the power $P_{u,t}^{inj}$ of 3.43 kW (Figure 7d). The highest demand value of 5.03 kW (Figure 3c) in this period is related to prosumer 3. And in this circumstance, $P_{u,t}^{s}$ adopts a positive value of 5.52 kW (Figure 6f). In this way, the SB absorbs approximately 9% of $P_{u,t}^{s}$ (Figure 7f). This fact shows the influence of the power supplied by the electric network and the operation of the SBs to serve each smart home during the peak period.

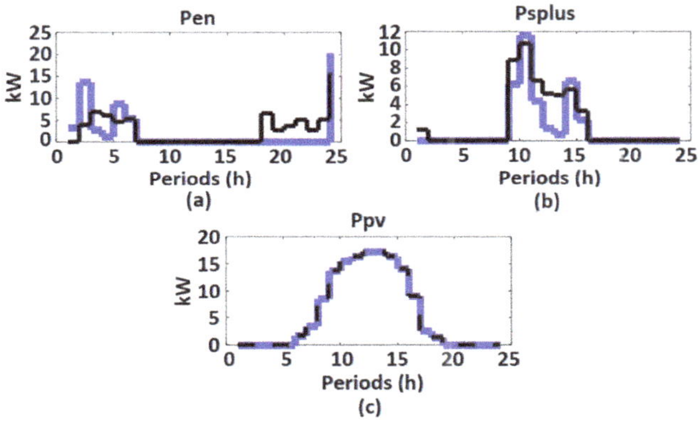


**Figure 5.** Variation during the day of the (**a**) power supplied by the energy company, (**b**) surplus power injected into the power grid, and (**c**) power produced by the shared PV plant.

Figures 7 and 8 show the injection (negative values) or absorption (positive values) of power, as well as the state of charge of the SBs in the community of prosumers, respectively. In Figure 7, for both cases, the largest number of periods $t$ in which the SBs absorb power is within hours with the presence of solar radiation $\mathcal{G}_t$. As shown in Figure 5a, the power grid does not contribute to the community's energy needs during this time. Therefore, the PV plant, in addition to contributing to surplus power and the supply of household appliances, also assists the SBs within these hours. It is worth noting that, for Case 1, during peak hours, both the shared PV plant and the electricity grid do not supply energy to the community. Also, during this period, the total energy consumption of the community is zero (Figure 3d). Therefore, the injected power $P_{u,t}^{inj}$ and absorption power $P_{u,t}^{abs}$ of the SBs during the peak period shown in Figure 7a,c,e indicate the exchange of power between the same storage technologies. In Case 2, during peak hours, the SBs inject power with a maximum value of 3.0 kW (prosumer 1), 3.5 kW (prosumer 2), and 2.0 kW (prosumer 3). These power values allow managing the state of charge of each SB while the energy consumption of each prosumer is met. Finally, Figure 8 shows the state of charge, $SoC_{u,t}^{sb}$, of each SB for both cases. It is worth mentioning that the state of charge of each SB in the last period of the previous day ($t = 0$) is equal to 35%, 45%, and 25% of its capacity. In all cases, the SBs reach full charge (12 kWh for prosumer 1; 10 kWh for prosumer 2; and 15 kWh for prosumer 3) at the end of the day. In the operation of each SB, for Cases 1 and 2, its state of charge at $t = 1$ shows a reduction of the stored energy to a value of 10% of the capacity at the same time as the powers of 3.0 kW, 3.50 kW, and 2.5 kW are injected. Note that at the beginning of the day, i.e., from $t \geq 1$, the SBs do not have consecutive recharges. Also, note that consecutive recharge occurs within the hours with available solar radiation. In addition, in most cases, the SBs reach full recharge in periods close to the end of the afternoon. Thus, at night hours, when energy prices are high and there is no production of PV energy, SBs contribute to share power among themselves (Case 1) or meet the consumption of household appliances (Case 2).

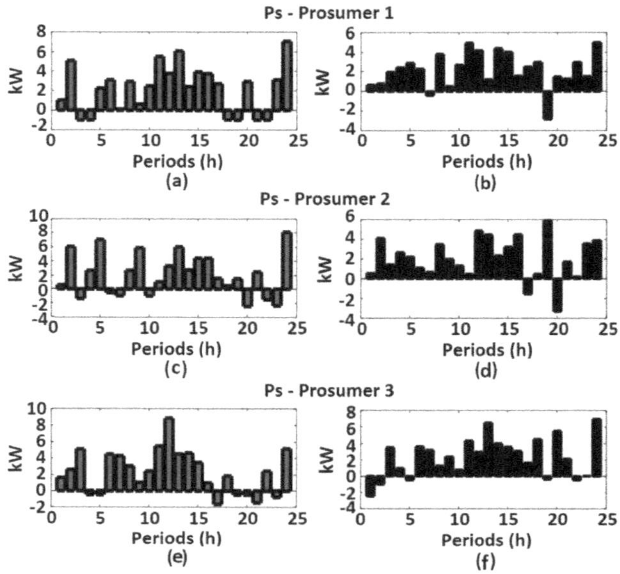


**Figure 6.** Bidirectional power $P^{s}_{u,t}$ measured by smart meter during the day for prosumer 1 ((**a**) Case 1 and (**b**) Case 2); prosumer 2 ((**c**) Case 1, and (**d**) Case 2); and prosumer 3 ((**e**) Case 1, and (**f**) Case 2).

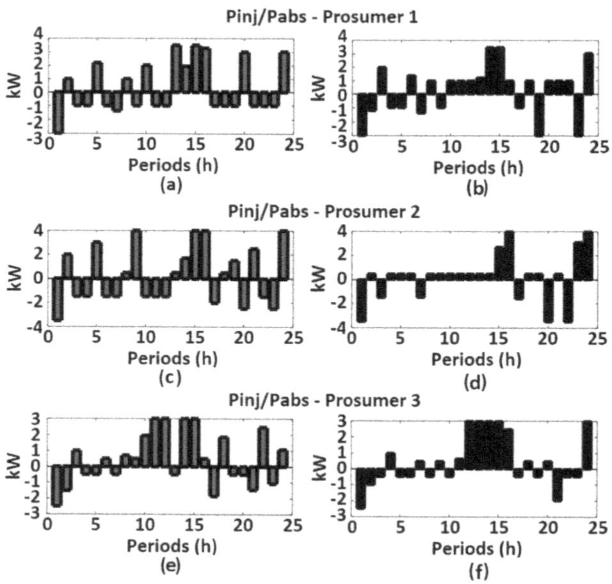


**Figure 7.** Injection $P^{inj}_{u,t}$ and absorption $P^{abs}_{u,t}$ powers related to the SB of the prosumer 1 ((**a**) Case 1 and (**b**) Case 2); prosumer 2 ((**c**) Case 1, and (**d**) Case 2); and prosumer 3 ((**e**) Case 1, and (**f**) Case 2).

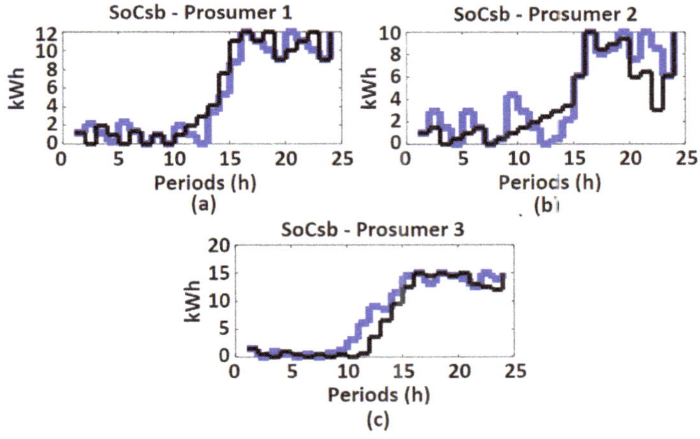


**Figure 8.** State of charge $SoC_{u,t}^{sb}$ of SBs for Case 1 (in blue) and Case 2 (in black) related to (**a**) prosumer 1; (**b**) prosumer 2; and (**c**) prosumer 3.

## 5. Conclusions

In this paper, a MILP model has been proposed to address the problem of intelligent energy management in a community of prosumers by improving the LF. The formulation of the proposed model considers a set of economic and operational constraints related to the household appliances usage and charging of EV batteries and the SBs present in each smart home. Constraints related to energy production by a shared PV plant, surplus energy, and community energy balance are also taken into account. Moreover, the efficient scheduling of appliances has been carried out considering the flexibility of their hours of usage, i.e., hourly preferences, without jeopardizing the prosumers' lifestyle. Uncertainties in the habitual usage of each appliance during the day were simulated using a Monte Carlo algorithm. In evaluating the effectiveness of the proposed model, two cases have been considered. In Case 1, the MILP model performs intelligent management without considering the LF improvement. For Case 2, the LF improvement is considered in energy management. The results have shown that the strategy that adopts the improvement of the LF guarantees financial benefits for both the energy companies and prosumers. For prosumers, this strategy allows minimizing energy consumption expenditures through the optimal scheduling of appliances, especially those with higher average power, to avoid the appearance of new consumption peaks during off-peak hours. Moreover, the strategy performs intelligent management of PV generation and energy storage technologies to meet the community's energy needs while any surplus energy produced is injected into the electricity grid. On the other hand, energy companies obtain financial benefits when the wide load distribution (i.e., increasing the LF) in the optimal consumption profile relieves the stress of the power grid and reduces energy waste during power supply. Consequently, several operating costs are minimized in the short and long term, among them the costs related to the maintenance of assets, e.g., power transformers, feeders, and protection devices.

**Author Contributions:** Investigation, F.V.C. and L.S.S.P.; methodology, F.V.C.; writing—review and editing, F.V.C., M.P.-K., E.N., and J.C.; resources, M.P.-K. and M.L. All authors have read and agreed to the published version of the manuscript.

**Funding:** The APC was funded by Aalto University.

**Institutional Review Board Statement:** Not applicable.

**Informed Consent Statement:** Not applicable.

**Data Availability Statement:** Not applicable.

**Conflicts of Interest:** The authors declare no conflict of interest.

## Nomenclature

| | |
|---|---|
| Functions | |
| $\psi_1$ | Cost function related to the optimal consumption profile of prosumers, appliances with higher average power, and the surplus power to be injected into the electric network of the energy company. |
| $\psi_2$ | Cost function related to the load factor of prosumers. |
| Indexes | |
| $u$ | Index for prosumers. |
| $a$ | Index for home appliances. |
| $t$ | Index for periods. |
| $y$ | Index for discrete blocks. |
| Sets | |
| $U$ | Set of prosumers $u$ |
| $A$ | Set of home appliances $a$ |
| $T$ | Set of periods $t$ |
| $Y$ | Set of discrete blocks $y$ |
| Parameters | |
| $\zeta_t$ | Energy price in period $t$ [\$/kWh]. |
| $P_a$ | Average power of appliance $a$ [kW]. |
| $\beta_a$ | Represents the type of appliance $a$: $-1$: EV; 0: appliance $a$ with working hours greater than or equal to 1 h; and 1: appliance $a$ with working hours less than 1 h. |
| $\beta_a^i$ | Binary parameter that adopts 1 for appliances with power higher than average. Otherwise, adopts 0. |
| $\mathfrak{C}_{a,t}$ | Usage probability of a given appliance $a$ in period $t$. |
| $\hat{\mathfrak{C}}_{a,t}$ | Accumulated probability related to the usage of a given appliance $a$ in period $t$. |
| $\Delta t$ | Time duration of each period $t$ [h]. |
| $\tau_a$ | Average value of usage time for the appliance $a$ [h]. |
| $\underline{\mathcal{Q}}_a$ | Minimum number of times that appliance $a$ with $\beta_a = 1$ is utilized. |
| $\overline{\mathcal{Q}}_a$ | Maximum number of times that appliance $a$ with $\beta_a = 1$ is utilized. |
| $\underline{\tau}_a$ | Minimum usage time of appliance $a$ with $\beta_a = 1$ [h]. |
| $\overline{\tau}_a$ | Maximum usage time of appliance $a$ with $\beta_a = 1$ [h]. |
| $\mathcal{Q}_a$ | Average value related to the number of times the appliance $a$ with $\beta_a = 0$ [kW]. |
| $\chi_{u,a,t}^{hp}$ | Binary matrix related to $\mathcal{H}_{u,a,t}^{cp}$. Indicates for each prosumer $u$, the usage state of appliance a in each period $t$. |
| $\mathcal{H}_{u,a,t}^{cp}$ | Continuous values matrix. Indicates for each prosumer $u$, the habitual energy consumption of each appliance $a$ in period $t$ [kWh]. |
| $M$ | Big value related to the linearization process. |
| $P^{ev}$ | EV charging rate [kW]. |
| $\underline{\tau}_u^{ev}$ | Minimum charging time of the EV related to prosumer $u$ [h]. |
| $\overline{\tau}_u^{ev}$ | Maximum charging time of the EV related to prosumer $u$ [h]. |
| $\underline{\mathcal{Q}}_u^{ev}$ | Minimum number of times the battery of the EV related to prosumer $u$ can be charged. |
| $\overline{\mathcal{Q}}_u^{ev}$ | Maximum number of times the battery of the EV related to prosumer $u$ can be charged. |
| $SoC_u$ | Initial state of charge related to EV battery of prosumer $u$ [kWh]. |
| $\mathcal{C}_u^{ev}$ | Energy storage capacity of the EV battery related to prosumer $u$ [kWh]. |
| $\lambda^1$ | Percentage value related to $\mathcal{C}_u^{ev}$. |
| $\theta_{u,a,t}^p$ | Hourly preferences. Indicates flexibility in the periods $t$ when prosumer $u$ can usage the home appliance $a$ without creating discomfort. |
| $\overline{\Delta}_t$ | Maximum value related to the variable $\Delta\Gamma_{t,y}$. |
| $\Pi_{t,y}$ | Inclination value related to the discrete block $y$ at period $t$. |
| $\underline{P}_u^{abs}$ | Minimum value of power absorbed by the SB related to prosumer $u$ [kW]. |
| $\overline{P}_u^{abs}$ | Maximum value of power absorbed by the SB related to prosumer $u$ [kW]. |
| $\underline{P}_u^{inj}$ | Minimum value of power injected by the storage battery related to prosumer $u$ [kW]. |
| $\overline{P}_u^{inj}$ | Maximum value of power injected by the SB related to prosumer $u$ [kW]. |
| $\eta_u^{abs}$ | Efficiency in power absorption by the SB related to prosumer $u$. |
| $\eta_u^{inj}$ | Efficiency in power injection by the SB related to prosumer $u$. |
| $\mathcal{C}_u^{sb}$ | SB capacity related to prosumer $u$ [kWh]. |
| $\lambda_u^2$ | Percentage value related to $\mathcal{C}_u^{sb}$. |
| $P_t^{pv}$ | Power supplied by the photovoltaic plant in each period $t$ [kW]. |
| k | Accumulator. |

| | |
|---|---|
| $\hat{\mathcal{G}}$ | Maximum value of the local solar radiation [kW/m$^2$]. |
| $\mathcal{G}_t$ | Local solar radiation in each period $t$ [kW/m$^2$]. |
| $\gamma_t$ | Standard solar radiation profile per unit. |
| $\mathcal{A}^{pv}$ | Photovoltaic area [m$^2$]. |
| $\zeta^p$ | Reduction factor due to production tolerance. |
| $\zeta^{ti}$ | Reduction factor due to temperature increase. |
| $\zeta^{dd}$ | Reduction factor due to the presence of dirt and dust. |
| $\zeta^{ml}$ | Reduction factor due to mismatch and wiring losses. |
| $\zeta^{cl}$ | Reduction factor due to DC to AC conversion losses. |
| $\varpi^{(1)}$ | Weighted weight related to the first component of function $\psi_1$. |
| $\varpi^{(2)}$ | Weighted weight related to the second component of function $\psi_1$. |
| $\varpi^{(3)}$ | Weighted weight related to the third component of function $\psi_1$. |
| $\varpi^{(4)}$ | Weighted weight related to single component of function $\psi_2$. |
| Variables | |
| $\mathrm{X}^{op}_{u,a,t}$ | Binary matrix related to $\mathcal{O}^{cp}_{u,a,t}$. Indicates for each prosumer $u$, the usage state of the appliance $a$ in each period $t$. |
| $\mathcal{O}^{cp}_{u,a,t}$ | Continuous values matrix. Indicates for each prosumer $u$, the optimal energy consumption of each appliance $a$ in period $t$ [kWh]. |
| $\tau^{us}_{u,a,t}$ | Represents for the prosumer $u$, the time of usage of the appliance $a$ in period $t$ [h]. |
| $\theta^{i}_{u,t}$ | Coincidence factor. Indicates for the prosumer $u$, the number of appliances that are utilized at the same period $t$. |
| $SoC^{ev}_{u,t}$ | Indicates for the prosumer $u$, the state of charge of the EV battery in each period $t$ [kWh]. |
| $\tau^{ev}_{u,t}$ | Represents for the prosumer $u$, the EV battery charging time in period $t$ [h]. |
| $\varepsilon^{ev}_{u}$ | Indicates for the prosumer $u$, the total energy stored in the EV battery [kWh]. |
| $\Delta\tau^{us}_{u,a,t}$ | Linearization variable related to $\mathrm{X}^{op}_{u,a,t} \times \tau^{us}_{u,a,t}$. |
| $\Delta\tau^{ev}_{u,a,t}$ | Linearization variable related to $\mathrm{X}^{op}_{u,a,t} \times \tau^{ev}_{u,t}$. |
| $P^{at}_t$ | Indicates for each period $t$, the power related to the total number of prosumers [kW]. |
| $P^{av}$ | Average value of $P^{at}_t$ [kW]. |
| $\Lambda_t$ | Represents the difference between $P^{at}_t$ and $P^{av}$ at period $t$ [kW]. |
| $\Lambda^+_t$ | Auxiliary variable to be used in the objective function discretization process. |
| $\Lambda^-_t$ | Auxiliary variable to be used in the objective function discretization process. |
| $\Delta\Lambda_{t,y}$ | Auxiliary variable to be used in the square of $\Gamma_t$ discretization process. |
| $P^{en}_t$ | Power supplied by the electricity distribution company in each period $t$ [kW]. |
| $P^{splus}_t$ | Surplus power sent to the electricity distribution network in each period $t$ [kW]. |
| $P^{s}_{u,t}$ | Indicate for the prosumer $u$, the bidirectional power measured by the smart meter in each period $t$ [kW]. |
| $P^{sr}_{u,t}$ | Power injected by SB related to the prosumer $u$ in period $t$ [kW]. |
| $P^{inj}_{u,t}$ | Indicates for the prosumer $u$, the power injected in each period $t$ [kW]. |
| $P^{abs}_{u,t}$ | Indicates for the prosumer $u$, the power absorbed in each period $t$ [kW]. |
| $\mu^{inj}_{u,t}$ | Binary variable that determines for the prosumer $u$ the injection status of the SB in each period $t$. |
| $\mu^{abs}_{u,t}$ | Binary variable that determines for the prosumer $u$ the absorption status of the SB in each period $t$. |
| $SoC^{sb}_{u,t}$ | Indicates for the prosumer $u$, the state of charge of the SB in each period $t$ [kWh]. |

## References


1. Rawlings, J.; Coker, P.; Doak, J.; Burfoot, B. Do smart grids offer a new incentive for SME carbon reduction? *Sustain. Cities Soc.* **2014**, *10*, 245–250. [CrossRef]
2. Abujubbeh, M.; Al-Turjman, F.; Fahrioglu, M. Software-defined wireless sensor networks in smart grid: An overview. *Sustain. Cities Soc.* **2019**, *51*, 101754. [CrossRef]
3. Joseph, A.; Balachandra, P. Energy internet, the future electricity system: Overview, concept, model structure, and mechanism. *Energies* **2020**, *13*, 4242. [CrossRef]
4. Yu, D.; Xu, X.; Dong, M.; Nojavan, S.; Jermsittiparsert, K.; Abdollahi, A.; Allah Aalami, H.; Pashaei-Didani, H. Modeling and prioritizing dynamic demand response programs in the electricity markets. *Sustain. Cities Soc.* **2020**, *53*, 101921. [CrossRef]
5. Pourakbari-Kasmaei, M.; Asensio, M.; Lehtonen, M.; Contreras, J. Trilateral Planning Model for Integrated Community Energy Systems and PV-Based Prosumers—A Bilevel Stochastic Programming Approach. *IEEE Trans. Power Syst.* **2020**. [CrossRef]
6. Miller, W.; Senadeera, M. Social transition from energy consumers to prosumers: Rethinking the purpose and functionality of eco-feedback technologies. *Sustain. Cities Soc.* **2017**, *35*, 615–625. [CrossRef]
7. Strielkowski, W.; Streimikiene, D.; Fomina, A.; Semenova, E. Internet of energy (IoE) and high-renewables eletricity systems market design. *Energies* **2019**, *12*, 4790. [CrossRef]
8. Espe, E.; Potdar, V.; Chang, E. Prosumer communities and relationships in smart grids: A literature review, evolution and future directions. *Energies* **2018**, *11*, 2528. [CrossRef]
9. Li, L.; Yu, S. Optimal management of multi-stakeholder distributed energy systems in low-carbon communities considering demand response resources ans carbon tax. *Sustain. Cities Soc.* **2020**, *61*, 102230. [CrossRef]
10. Guichi, A.; Talha, A.; Madjid Berkouk, E.; Mekhilef, S. Energy management and performance evaluation of grid connected PV-battery hybrid system with inherent control scheme. *Sustain. Cities Soc.* **2018**, *41*, 490–504. [CrossRef]

11. Tascikaraoglu, A. Economic and operational benefits of energy storage sharing for a neighborhood of prosumers in a dynamic pricing environment. *Sustain. Cities Soc.* **2018**, *38*, 219–229. [CrossRef]
12. Rahman, M.; Arefi, A.; Shafiullah, G.; Hettiwatte, S. A new approach to voltage management in unbalanced low voltage network using demand response and OLTC considering consumer preference. *Int. J. Elec. Power.* **2018**, *99*, 11–27. [CrossRef]
13. Morteza Ghorashi, S.; Rastergar, M.; Senemmar, S.; Reza Seifi, A. Optimal design of reward-penalty demand response programs in smart power grids. *Sustain. Cities Soc.* **2020**, *60*, 102150. [CrossRef]
14. European Comission–Gfk Belgium Consortium. Study on Residential Prosumers in the European Energy Union. Available online: https://ec.europa.eu/commission/sites/beta-political/files/study-residential-prosumers-energy-union_en.pdf (accessed on 23 December 2020).
15. Nuchprayoon, S. Calculation and allocation of load losses in distribution system using load research data and load factor method. In Proceedings of the 6th IEEE International Conference on Control System, Computing and Engineering (ICCSCE), Batu Ferringhi, Malaysia, 25–27 November 2016; pp. 85–90.
16. Saikira, B.; Manas, M.; Baruah, D. Distribution loss reduction in a university of north east india through load factor improvement. In Proceedings of the 2015 International Conference on Energy Systems and Applications, Pune, India, 30 October–1 November 2015; pp. 203–208.
17. Surai, J.; Surapatana, V. Load factor improvement in industrial sector load duration curves. In Proceedings of the 2014 International Electrical Engineering Congress (iEECON), Chonburi, Thailand, 19–21 March 2014; pp. 1–4.
18. Cerna, F.V.; Contreras, J. A MILP model to relieve the occurrence of new demand peaks by improving the load factor in smart homes. *Sustain. Cities Soc.* **2021**, *71*. [CrossRef]
19. Fourer, R.; Gay, D.M.; Kernighan, B.W. *AMPL: A Modeling Language for Mathematical Programming*, 2nd ed.; Duxbury Press/Brooks/Cole Publishing Company: Boston MA, USA, 2001; Available online: https://ampl.com/resources/the-ampl-book/ (accessed on 17 March 2020). [CrossRef]
20. IBM ILOG CPLEX Optimization Studio CPLEX User's Manual–Version 12 Release 8. Available online: https://www.ibm.com/support/knowledgecenter/SSSA5P_12.8.0/ilog.odms.studio.help/pdf/usrcplex.pdf (accessed on 20 October 2020).
21. AssLiu, X.; Wang, S.; Sun, J. Energy management for community energy network with CHP based on cooperative game. *Energies* **2018**, *11*, 1066. [CrossRef]
22. Radl, J.; Fleischhacker, A.; Huglen Revheim, F.; Lettner, G.; Auer, H. Comparison of profitability of PV electricity sharing in renewable energy communities in selected european countries. *Energies* **2020**, *13*, 2644. [CrossRef]
23. De la Hoz, J.; Alonso, A.; Coronas, S.; Martin, H.; Matas, J. Impact of different regulatory structures on the management of energy communities. *Energies* **2020**, *13*, 2892. [CrossRef]
24. Marangoni, F.; Magatão, L.; Ramos de Arruda, L. Demand response optimization model to energy and power expenses analysis and contract revision. *Energies* **2020**, *13*, 2803. [CrossRef]
25. Wang, J.; Garifi, K.; Baker, K.; Zou, W.; Zhang, Y.; Huang, S.; Vrabie, D. Optimal renewable resource allocation and load scheduling of resilient communities. *Energies* **2020**, *13*, 5683. [CrossRef]
26. Ghosh, A.; Aggarwai, V. Penalty based control mechanism for strategic prosumers in a distribution network. *Energies* **2020**, *13*, 452. [CrossRef]
27. Correa-Florez, C.; Michiorri, A.; Kariniotakis, G. Comparative analysis of adjustable robust optimization alternatives for the participation of aggregated residential prosumers in eletricity markets. *Energies* **2019**, *12*, 1019. [CrossRef]
28. Ur Rashid, M.; Granelli, F.; Hossain, A.; Alam, S.; Saleh Al-Ismail, F.; Kumar Karmaker, A.; Rahaman, M. Development of home energy management scheme for a smart grid community. *Energies* **2020**, *13*, 4288. [CrossRef]
29. Alabdullatif, A.; Gerding, E.; Perez Diaz, A. Market design and trading strategies for community energy markets with storage and renewable supply. *Energies* **2020**, *13*, 972. [CrossRef]
30. Li, L. Optimal coordination strategies for load service entity and community energy systems based on centralized and decentralized approaches. *Energies* **2020**, *13*, 3202. [CrossRef]
31. Bukhsh, R.; Umar Javed, M.; Fatima, A.; Javaid, N.; Shafiq, M.; Ghoo Choi, J. Cost efficient real time electricity management services for green community using fog. *Energies* **2020**, *13*, 3164. [CrossRef]
32. Kuruseelan, S.; Vaithilingam, C. Peer-to-peer energy trading of a community connected with an ac and dc microgrid. *Energies* **2019**, *12*, 3709. [CrossRef]
33. González-Romera, E.; Ruiz-Cortés, M.; Milanés-Montero, M.; Barrero-Gonzáles, F.; Romero-Cadaval, E.; Amaral Lopes, R. Martins, J. Advantages of minimizing energy exchange instead of energy cost in prosumer microgrids. *Energies* **2019**, *12*, 719. [CrossRef]
34. Shin, I. Approximation algorithm-based prosumer scheduling for microgrids. *Energies* **2020**, *13*, 5853. [CrossRef]
35. Achiluzzi, E.; Kobikrishna, K.; Sivabalan, A.; Sabillon, C.; Venkatesh, B. Optimal asset planning for prosumers considering energy storage and photovoltaic (PV) units: A stochastic approach. *Energies* **2020**, *13*, 1813. [CrossRef]
36. Toquica, D.; Agbossou, K.; Malhamé, R.; Henao, N.; Kelouwani, S.; Cardenas, A. Adaptive machine learning for automated modeling of residential prosumer agents. *Energies* **2020**, *13*, 2250. [CrossRef]
37. De Greve, Z.; Bottieau, J.; Vangulick, D.; Wautier, A.; David Dapoz, P.; Arrigo, A.; Toubeau, J.; Vallée, F. Machine learning techniques for improving self-consumption in renewable energy communities. *Energies* **2020**, *13*, 4892. [CrossRef]

38. García-Villalobos, J.; Zamora, I.; Eguia, P.; Torres, E.; Etxegarai, A.; San Martin, J.I. Optimization of load factor in distribution networks with high share of plug-in electric vehicles and photovoltaic generation. In Proceedings of the 2017 52nd International Universities Power Engineering Conference (UPEC), Heraklion, Greece, 28–31 August 2017; pp. 1–6.
39. Trongwanichnam, K.; Thitapars, S.; Leeprechanon, N. Impact of plug-in electrici vehicles load planning to load factor and total generation cost in a power system. In Proceedings of the 2019 IEEE PES GTD Grand international and Exposition Asia (GTD Asia), Bangkok, Thailand, 19–23 March 2019; pp. 599–604.
40. Mohg Ali, S.B.; Hasanuzzaman, A.; Rahim, N. Investigation on the load factor performance at Wisma R&D universiti Malaya building. In Proceedings of the 5th IET International Conference on Clean Energy and Technology (CEAT2018), Kuala Lumpur, Malaysia, 5–6 September 2018; pp. 1–6.
41. Chiu, W.Y.; Hsieh, J.T.; Chen, C.M. Pareto optimal demand response based on energy costs and load factor in smart grid. *IEEE Trans. Ind. Inform.* **2020**. [CrossRef]
42. Al Fardan, A.S.; Al Gahtani, K.S.; Asif, M. Demand side management solution through new tariff structure to minimize excessive load growth and improve system load factor by improving commercial buildings energy performance in Saudi Arabia. In Proceedings of the 2017 IEEE International Conference on Smart Energy Engineering (SEGE), Oshawa, ON, Canada, 14–17 August 2017; pp. 308–320.
43. Lorestani, A.; Aghaee, S.S.; Gharehpetian, G.B.; Ardehali, M.M. Energy management in smart home including PV panel, battery, electric heater with integration of plug-in electric vehicle. In Proceedings of the 2017 Smart Grid Conference (SGC), Tehran, Iran, 20–21 December 2017; pp. 1–7.
44. Cho, D.; Valenzuela, J. Optimization of residential off-grid PV-battery systems. *Sol. Energy* **2020**, *28*, 647–653. [CrossRef]
45. Liu, N.; Wang, J.; Yu, X.; Ma, L. Hybrid energy sharing for smart building cluster with CHP system and PV prosumers: A coalitional game approach. *IEEE Access* **2018**. [CrossRef]
46. Muqeet, H.A.U.; Ahmad, A. Optimal scheduling for campus prosumer microgrid considering price based demand response. *IEEE Access.* **2020**. [CrossRef]
47. Barnes, A.K.; Balda, J.C.; Hayes, J.K.; Modelling, P.V. Clouding Effects Using a Semi-Markov Process with Application to Energy Storage. *IFAC Proc. Vol.* **2014**, *47*, 9444–9449. [CrossRef]
48. Lappalainen, K.; Valkealahti, S. Effects of PV array layout, electrical configuration and geographic orientation on mismatch losses caused by moving clouds. *Solar Energy* **2017**, *144*, 548–555. [CrossRef]
49. Lappalainen, K.; Valkealahti, S. Photovoltaic mismatch losses caused by moving clouds *Solar Energy* **2017**, *158*, 455–561. [CrossRef]
50. Element Energy, Further Analysis of the Household Electricity Use Survey–Electrical Appliances at Home: Tuning in to Energy Saving. Available online: https://assets.publishing.service.gov.uk/government/uploads/system/uploads/attachment_data/file/275484/electricity_survey_2_tuning_in_to_energy_saving.pdf (accessed on 10 December 2020).
51. UNDP CEDRO Project, Energy Efficient Home Appliances: Perspctives from Lebanese Consumers. Available online: https://www.undp.org/content/dam/lebanon/docs/Energy%20and%20Environment/Publications/CEDRO%20_%20Energy%20Efficient%20Home%20Appliances.pdf (accessed on 15 November 2020).
52. Robert, P.; Casella, G. *Monte Carlo Statistical Methods*, 2nd ed.; Springer: New York, NY, USA, 2004; p. 157. [CrossRef]
53. US Department of Energy. Appliance & Equipment Standards/Standards and Test Procedures. Available online: https://www.energy.gov/eere/buildings/standards-and-test-procedures (accessed on 20 December 2020).
54. Data. Available online: https://drive.google.com/file/d/1i50a68nRO2zEl8pjPIwlOYKEmOWdMnhS/view?usp=sharing (accessed on 2 January 2021).
55. Yahia, Z.; Pradhan, A. Multi-objective optimization of household appliance scheduling problem considering consumer preference and peak load reduction. *Sustain. Cities Soc.* **2020**, *55*. [CrossRef]
56. Cerna, F.V.; Pourakbari-Kasmaei, M.; Lehtonen, M.; Contreras, J. Efficient automation of an HEV heterogeneous fleet using a two-stage methodology. *IEEE Trans. Veh. Technol.* **2019**. [CrossRef]
57. Borges, M.C.O.; Franco, J.F.; Rider, M.J. Optimal reconfiguration of electrical distribution systems using mathematical programming. *J. Control. Autom. Electr. Syst.* **2014**, *25*. [CrossRef]
58. Gonçalves, R.R.; Alves, R.P.; Franco, J.F. Operation planning of electrical distribution systems using a mixed integer linear model. *J. Control. Autom. Electr. Syst.* **2013**, *24*. [CrossRef]
59. National Electric Energy Agency (ANEEL), Normative Resolution n°. 733–White Tariff (2016). Available online: https://www.aneel.gov.br/tarifa-branca (accessed on 10 July 2020).
60. Binkley, A. Solar Technology Reference Guide. NAIOP Research Foundation. Available online: File:///C:/Users/Samsung/Downloads/solartechrefguide.pdf (accessed on 15 July 2020).
61. Goodrich, A.; James, T.; Woodhouse, M. Residential, Commercial, and Utility-Scale Photovoltaic (PV) Systems Prices in the United States: Current Drivers and Cost-Reduction Opportunities, Technical Report NREL/TP-6 A20-53347. Available online: https://www.nrel.gov/docs/fy12osti/53347.pdf (accessed on 25 October 2020).
62. Xie, H.; Teng, X.; Xu, Y.; Wang, Y. Optimal energy storage sizing for networked microgrids considering reliability and resilience. *IEEE Access.* **2019**. [CrossRef]
63. Asian Development Bank. *Handbook on Battery Energy Storage System*; 6 ADB: Mandaluyong City, Manila, Philippines, 2018; p. 157. [CrossRef]

64. Digital Atlas-Solar Radiation (W/m2)–April 2020. Available online: https://www.cnpma.embrapa.br/climapest/atlasdigital/radiacao_solar/2020/html/MED_RAD_SOLAR_2020_A2_ABR.html (accessed on 10 May 2020).
65. Melhem, F.Y.; Grunder, O.; Hammoudan, Z.; Moubayed, N. Energy management in electrical smart grid environment using robust optimization algorithm. *IEEE Trans. Ind. Appl.* **2018**. [CrossRef]
66. SUNTECH–330 Watt Polycrystalline Solar Module. Available online: http://www.get-systems.com/productsfiles/solarcells/AU_STP330_VfwMC4_330_325_320_1500V.pdf (accessed on 15 June 2020).

*energies*

MDPI

*Article*

# Implementation and Simulation of Real Load Shifting Scenarios Based on a Flexibility Price Market Strategy—The Italian Residential Sector as a Case Study

**Francesco Mancini [1,*], Jacopo Cimaglia [2], Gianluigi Lo Basso [3] and Sabrina Romano [4]**

[1] Department of Planning, Design and Technology of Architecture, Sapienza University of Rome, 72-00197 Rome, Italy
[2] Interdepartmental Center for Territory, Building, Conservation and Environment, Sapienza University of Rome, 53-00197 Rome, Italy; jacopocimaglia.ingegneria@gmail.com
[3] Department of Astronautics, Electrical Energy Engineering, Sapienza University of Rome, 18-00184 Rome, Italy; gianluigi.lobasso@uniroma1.it
[4] Energy Technologies Department (DTE), Italian National Agency for Technologies, Energy and Sustainable Economic Development (ENEA), 301-00123 Rome, Italy; sabrina.romano@enea.it
* Correspondence: francesco.mancini@uniroma1.it; Tel.: +39-06-4991-9172

**Abstract:** This work aims to evaluate the Flexibility Potential that a residential household can effectively provide to the public grid for participating in a Demand Response activity. In detail, by using 14 dwellings electrical data collection, an algorithm to simulate the Load Shifting activity over the daytime is implemented. That algorithm is applied to different scenarios having considered the addition of several technical constraints on the end users' devices. In such a way, more realistic demand-side management actions are implemented in order to assess the Flexibility Potential deriving from the loads shifting. Basically, by performing simulations it is possible to investigate how the household appliances real operating conditions can reduce the theoretical Flexibility Potential extent. Starting from a Flexibility Price-Market-based Strategy, this work simulates the shifting over the day and night-time of some flexible loads, i.e., the shiftable and the storable ones. Specifically, all instants where load curtailments and enhancements occur over the typical day, the flexibility strategy effectiveness in terms of percentage, the power and energy that are potentially flexible, are evaluated. All the simulations are performed only for residential consumers to evaluate the actual dwellings Flexibility Potential in the absence of any electrical storage and production systems. The outcomes of these simulations show an average Theoretical Flexibility reduction, which is calculated as the fraction of appliances' cycles shifting over the total ones, equal to 53%, instead of 66%; in a single dwelling, a maximum variation equal to 29% has been registered. In the end, the monthly average shifted energy per dwellings decreases from 27 to 18 kWh, entailing 32.5% off.

**Keywords:** residential users; demand response; flexible loads shifting scenarios

check for updates

**Citation:** Mancini, F.; Cimaglia, J.; Lo Basso, G.; Romano, S. Implementation and Simulation of Real Load Shifting Scenarios Based on a Flexibility Price Market Strategy—The Italian Residential Sector as a Case Study. *Energies* **2021**, *14*, 3080. https://doi.org/10.3390/en14113080

Academic Editors: Mahdi Pourakbari-Kasmaei and Seppo Sierla

Received: 4 May 2021
Accepted: 21 May 2021
Published: 25 May 2021



## 1. Introduction

The European Union long-term strategic vision concerns a detailed analysis of all those actions to be undertaken for a greenhouse gasses zero emissions economy within 2050. The outlined scenarios provide the use of Renewable Energy Sources (RES) on large-scale [1]. However, a large quantity of electricity produced by non-programmable RES, may cause electrical grid management problems, due to potential mismatch between energy supply and energy demand [2].

An efficient balancing method is represented by the Demand-Side Management activities (DSM), which have the goal of encouraging consumers to modify their electrical energy use, either reducing their consumptions or shifting the power uptakes towards off-peak hours; in detail, among the DSM activities, a great interest was addressed to the Demand Response (DR) [3], which aims at reshaping the users' power demand profiles according to the grid requirements. DR helps public service enterprises and users to reduce peak power demand and the energy prices variability [4], converting the users into load management market active participants [5].

From a literature overview, it emerged how the DSM strategies implementation was adopted very often to improve energy and economic savings in big industrial sites; Notwithstanding, in the recent years, an increasing attention to the potential role of residential sector in the flexibility market has been paid. Even though costly ICT infrastructures to send signals from or towards the individual house are required, several research activities were focused on developing less expensive components and sensors to make them accessible and handy in the medium short term. Additionally, potential incentive schemes have been proposed in order to improve the end-users' profitability and to contribute to a wider deployment.

The growing interest in the implementation of DR programs to dwellings is basically due to the high energy needs in the building sector; indeed, in the EU member states, the real estate energy needs accounted for 40.3% of total consumption; namely, the building stock alone accounted for 26.1% of total energy consumptions, in accordance with data referred to 2018 [6]. It is noteworthy that a part of residential consumption is electrical and only a fraction of can be considered appropriate to be used flexibly. Having said this, a great potentiality is represented by all those buildings equipped with electric heat generators; the positive effect magnitude on overall consumption is strongly related to the climatic zone (i.e., outdoor temperature and relative humidity), to the buildings' envelope as well as to the occupancy rate [7].

Several research projects addressed the issue associated to the identification of the most proper methodology to assess the buildings' potential of flexibility [8]. For instance the authors in Ref. [9] proposed a predictive model to accurately schedule both the users and the energy resources which can be deferred over the day. The buildings' thermal mass can be reputed as a potential storage medium [10,11]. Indeed, that mass, which is a specific feature of each dwelling, it can store a certain amount of heat by either postponing or anticipating the operating time schedule of heating and cooling systems. In such a way, the indoor thermal comfort conditions can be kept under control to the standard set point [12]. Among the different available options to handle the load flexibility, the so-called Power-to-X strategies are currently offering good perspectives. Usually, the X letter is used as an umbrella terms to indicate the electricity conversion, hailing from the renewables' overcapacity, into different useful energy forms. Referring to the building sector, Power-to-Power, Power-to-Heat and Power-to-Gas are considered, by the scientific community, as the most promising and suitable options in the short-medium term [13–16]. However, cheaper and reliable storage devices, such as PCMs (Phase-change Materials) pressurized gas vessels, batteries and the electro-fuels injection into NG pipelines, have to be effectively embedded within the existing energy scenarios [17]. On the other hand, the recent literature, dealing with how to conveniently store heat in the residential sector, is strongly oriented to analyse the widespread electric heat pumps application for serving the end users in terms of Domestic Hot Water production (DHW as well [18]).

In accordance with the mentioned research indications, and based on what has emerged when applying the management strategy proposed in a previous work [19] different scenarios have been built. Specifically, they also include several constraints, in order to evaluate their effects on the dwellings' flexible potential. The study is based on a measurement campaign carried out over two years (2018–2019) related to 14 sample dwellings representing the middle regions Italian building stock; that cluster was chosen by selecting the most frequent typologies within a database consisting of 751 real dwellings. The database was built over the last three years by collecting a wide group of information such as geometrical characteristics, building materials and energy bills.

Having said this, the authors deem that their contribution to the knowledge in this research topic is substantially the methodological approach, which integrates on-field measurements with a simulation process; more specifically, the authors attempt to identify how the DSM strategies effectiveness drift away from theoretical values once realistic technical constraints on devices operation have been considered in calculation model. By combining several tools for statistical analysis, the dwellings' electrical loads have been analysed, classified, filtered, and then processed to verify the realistic flexibility. Moreover, this study aims at contributing to identify the potential role of Italian residential sector within the long-term strategy of progressively transforming the end users' energy consumptions. Indeed, a high electrification degree is well recognised as one of the most important drivers in the energy transition, towards an efficient RES integration within the national grid.

## 2. Materials and Methods

An effective DR program implementation generally needs an accurate knowledge of what a generic dwelling can offer, in terms of shiftable loads, for participating in such a strategy. Considering a single household, the flexibility potential is low and discontinuous very often; nevertheless, by gathering several dwellings it is possible to mitigate such discontinuities and to flatten the flexible loads amount over specific time spans. In so doing, building clusters or districts can fruitfully contribute to the electrical systems management along with improving their own safety and reliability issues [20].

The loads classification based on their intrinsic nature [21] can be useful only for preliminary analysis (see Figure 1), providing to designers with a flexibility amount overestimation: (i) all those loads deemed as flexible in accordance with the classification might be not-flexible related to the specific management strategy [22,23]; (ii) other loads might be not-flexible caused by different technical constraints related to the household appliances [24]. Those ones, can be eventually imposed by a building energy management system (e.g., BEMS) [25]. This latter, including several probes, sensors, transducers and interfaces, is characterised by a purchase price ranging between 1000 and 4000 € as the household size increases together with technical requirements, specifications and remote controls [26].

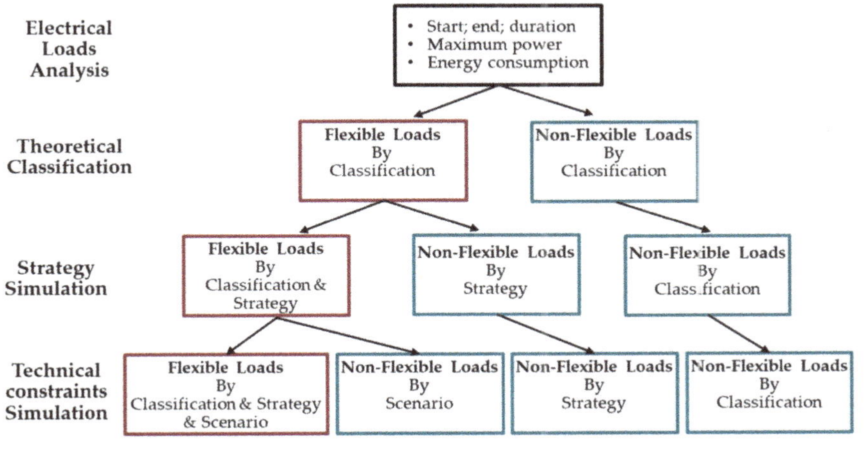


**Figure 1.** Flexible Loads quantification process.

In this work, the management strategy developed in a previous project of the same authors [19], are considered; this strategy was defined comparing the power demand profiles of a buildings cluster with the Italian hourly electricity price. In so doing, all those moments over the day, in which clusters should reduce or enhance their electrical uptakes, were detected. Consequently, optimized and reshaped power demand profiles, sorted by months and days typology, were built (i.e., weekdays, Saturdays, non-working days), as reported in Tables A1–A3 of Appendix A.

Seven technical constraints have been identified (see Table 1) and they can be outlined as follows: (C1) is a maximum flexibility window of 24 h [27]; (C2) is the maximum withdrawable power from the grid as a function of delivery contracts related to the end user typology; (C3) the dwelling occupancy for some appliances operation (vacuum cleaner, iron, etc.) [28]; (C4, C5) account for the working sequence of different appliances (washing machine, tumble dryer, dishwasher); (C6) takes into account the correlation between the users stochastic behaviour and the heating and cooling systems operation [29]; (C7) refers to different user settings imposed to avoid the satisfaction level lessening (of users themself and neighbours), e.g., for night-time noises.

**Table 1.** Operative bonds to the appliances working.

| Constraints (C) | Criterium | Definition |
|---|---|---|
| C1 | Flexibility window | Maximum shifting within 24 h ahead |
| C2 | Maximum power at the meter (detachment conditions) | P > 14.0 kW for $\tau$ > 2 s<br>P > 4.2 kW for $\tau$ > 2 min<br>P > 3.3 kW for $\tau$ > 182 min |
| C3 | Vacuum cleaner and Iron using | Occupancy in dwelling |
| C4 | Tumble Dryer and Washing Machine | TD operation within 3 h from WM cycle end |
| C5 | Dish Washer | End of operation within next meal |
| C6 | Heating and Cooling | Occupancy within the next 4 h (i.e., switching on within the previous 4 h from the original starting) |
| C7 | No noise in the night-time | No appliances shifting towards night-time between 12:00 a.m. and 06:00 a.m. |

Starting from the theoretical classification (S0), four scenarios were simulated with the aim of quantifying the flexible loads in the dwellings. Each scenario is characterised by the application of the aforementioned management strategy, along with the different constraints set and by the different impositions of the bond conditions set out above (see Table 2). The simulations were implemented using the Excel environment, with Macros written in Visual Basic for Applications (VBA).

**Table 2.** Simulated scenarios.

| Scenario (S) | Criterium |
|---|---|
| S0 | Theoretical Classification |
| S1 | Load Shifting Strategy Simulation;<br>No constraint Applied |
| S2 | Load Shifting Strategy Simulation;<br>constraints V1, V2, V3 Applied |
| S3 | Load Shifting Strategy Simulation;<br>constraints V1, V2, V3, V4, V5, V6 Applied |
| S4 | Load Shifting Strategy Simulation;<br>constraints V1, V2, V3, V4, V5, V6, V7 Applied |

The theoretical classification (S0) is based on a preliminary analysis carried out on the sample dwellings. It allows to estimate the flexibility in a "steady" way, once all of the flexible appliance cycles have been considered shiftable only by their nature; that implicitly means families are totally willing to participate in a DR program [21].

As regards the management strategy (S1) application, general flexible cycle hourly allocations and load shifting requests (i.e., "Load Reduction Time"; "Load Increase Time") were analysed and correlated, according to what is reported in Appendix A, Tables A1–A3);

as a result, some loads have been considered not-flexible. In detail, loads belonging to that category are all those off-takes occurring at the right hourly allocation (cycle load in "Load Increase Time" o "No Load Variation"), together with the non-shiftable ones due to the lack of "Load Increase Time" subsequent requests.

The addition of further constraints, such as S2, S3 and S4, limits load amount to be shifted, owing to the incompatibility between some hourly allocation "Load Increase Time" and scenario's constraints.

The flexible loads were characterised by considering, for each working cycle of any appliances, the following parameters: (i) the starting time, (ii) the cycle duration; (iii) the consumed energy ($E_{Flex}$); (iv) the maximum power ($P_{max}$).

The strategy and constraints application effects were evaluated by several indicators able to describe and summarize the flexibility amount in terms of statistical and energy values:

- *Real Flexibility (RF)* is the effective fraction of the executed load shifting, respect to the appliance total cycles tally (see Equation (1));
- *Energy Shift (ES)* is the shifted energy consumption deriving from the adopted actions (see Equation (2));
- *Peak Shaving (PS)* is the maximum achievable peak reduction by the load shifting, in terms of power, between the scenario Sx and the theoretical one without shifting (S0) (see Equation (3)).

$$RF_{\text{Sx}} = \frac{N_{Flex,\text{Sx}}}{N_{flex,\,\text{S0}}}. \tag{1}$$

$$ES_{\text{Sx}} = \sum E_{Flex,\text{Sx}} \tag{2}$$

$$PS_{\text{Sx}} = Max(P_{max,\text{S0}} - P_{max,\text{Sx}}) \tag{3}$$

where:

- $N_{Flex,\text{S0}}$ is the total tally of flexible cycles before applying both strategy and constraints (i.e., it corresponds to scenario S0);
- $N_{Flex,\text{Sx}}$ represents the number of effective executed load shifting, due to the scenario Sx application;
- $P_{max,\text{S0}}$ . indicates the maximum registered power (in the time span of 15 min) before applying the strategy and constraints (i.e., scenario S0);
- $P_{max,\text{Sx}}$ represents the maximum registered power (in the time span of 15 min), hailing from scenario Sx application, at the same time in which the $P_{max,\text{S0}}$ value occurs.

The first indicator (RF) was expressly defined for the present analysis; the remaining two (ES, PS) were widely used in literature to evaluate the flexibility in the residential sector and to define the political implications of different scenarios in the energy markets [30]. They were also combined with the cost reduction assessment [31] and they were included for defining power storage management strategies [32].

The explained procedure was applied to 14 sample dwellings of 751, selected by the use of a categorizing algorithm [19]. Therefore, they are considered as archetypes representing part of the Italian residential sector and their main features are outlined in Appendix B. In such dwellings, some sensors were installed to monitor the electrical consumption (Appendix B, Table A4) and a measurement campaign was carried out over 2018–2019.

The methodology application has been preceded by data elaboration process of the acquired measurement. Indeed, the sensors sampling time and data collection is about 5 s. The post-processing phase was performed in order to calculate average values over 15 min, according to the common Distributor's energy meters: Thus, all of calculations and analyses were carried out with these average values.

## 3. Results and Discussions

### 3.1. Flexible Loads by Classification

Using a data collection questionnaire [21], the required information for simulating dwellings energy consumptions and for classifying loads were obtained. Specifically collected data refer to the physical and geometrical dwelling characteristics, to plant equipment, to building occupancy, to household appliances typology and to their utilization (Appendix B, Tables A5 and A6). However, the selected archetypes show a modest electrification degree, equal to 36.8%; furthermore, all of samples have a NG-based heating system; only one dwelling shows a DHW production by heat pump system; in 9 dwellings cooling system are installed for the air conditioning of some rooms (2 units per dwelling on average). All the dwellings are equipped with washing machine; a dishwasher is installed only in 11 dwellings, while the tumble dryer only in 4 dwellings; there are also some not-flexible appliances, such as ICTs, personal care items, refrigerators, ovens and kitchen appliances and others with a marginal usage. Among those, the vacuum cleaner and the iron are available in all dwellings and in 11 of them, respectively.

In the selected archetypes live different family typologies, in terms of composition to adequately represent the Italian residential sector. The distribution can be summarised as follows: 3 households composed by 2 kids, a working parent and a non-working one 2 households composed by 2 kids and 2 working parents; 4 households composed by 1 kid and 2 working parents; 2 households composed by 2 workers; one household composed by a single worker; 2 households composed by 2 non-workers.

Based on the data collection, some preliminary analyses were performed in order to identify the appliance cycles main features and the household's habits. In detail, a trend for the most energy-intensive appliances, characterised by the longest operation times was detected: (i) the dishwasher shows an average starting operation time, in all cases and over the all months, occurring at 5:30 p.m. with an average variance of +4:30 h (40th percentile) and −1:30 h (60th percentile); (ii) the washing machine is commonly switched on close to 2:15 p.m. with an average variance of +1:30 h (40th percentile) and −1:45 h (60th percentile); the air conditioner usage is very diversified between the dwellings (different duration cycles, different temperature set-points) with an average starting operation time around 5:15 p.m. and an average variance of ± 4:15 h. The remaining appliances show a higher variability and a lower cyclic nature. Therefore, it is not easy to identify a unique trend, exception for a large frequent time span occurring between 11:00 a.m. and 5:00 p.m. Anyway, it is clear how the users' working habits strongly affect appliance usage. Indeed the electrical loads are mostly concentrated between 5:00 p.m. and 10:00 p.m., since the variance from the average value is greatly shifted towards the early evening hours For instance, air conditioners and other minor appliances are dependent on the house occupancy; dishwashers must clean up dishes for dinner; washing machines and tumble dryers, if any, typically operate in the evening. Consequently, the power demand profiles related to the archetypes show peak values close to these hours, consistently with the national power demand profile [19].

The annual average daily profile of flexible loads related to the archetypes are depicted in Figures 2–4. It can be noticed how those loads are generally concentrated in the afternoon exception for the weekdays (see Figure 2), where they are available early in the morning or after dinner.

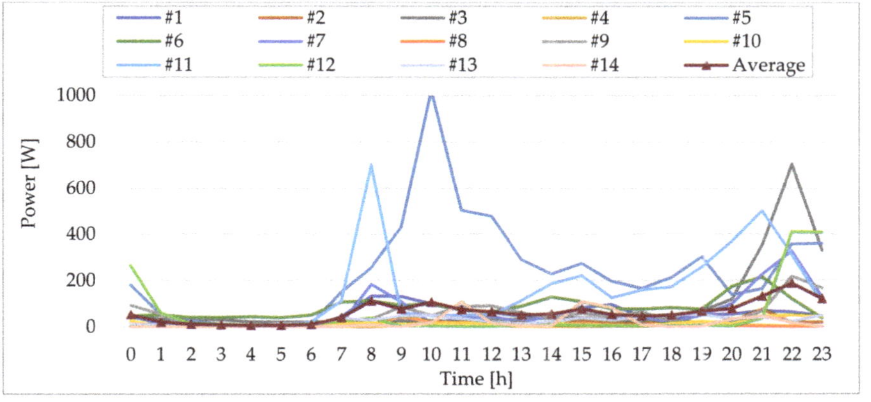


**Figure 2.** Archetypes average daily flexible loads profiles over the weekdays.

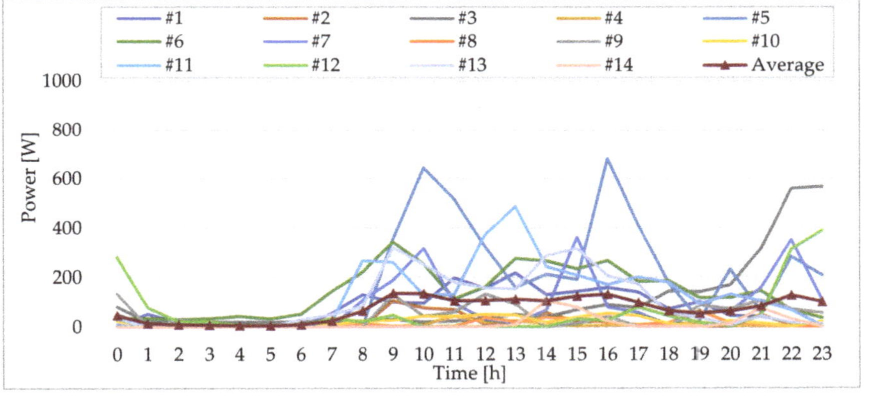


**Figure 3.** Archetypes average daily flexible loads profiles over the Saturdays.

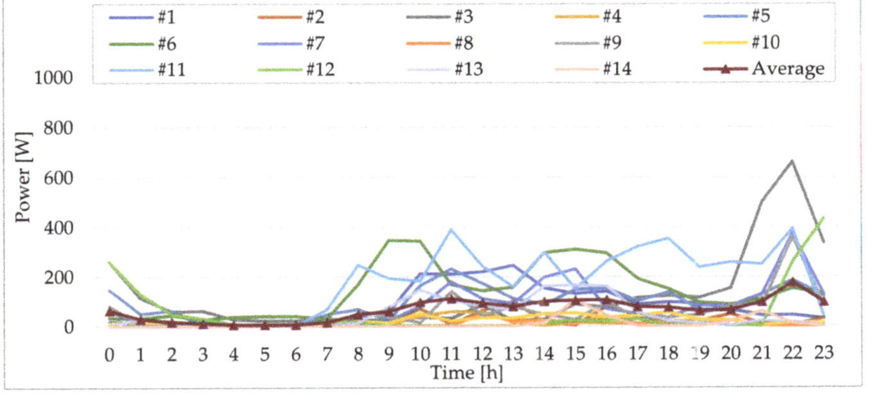


**Figure 4.** Archetypes average daily flexible loads profiles over non-working days.

Referring to plotted data in Figures 3 and 4, it emerges how each archetype is characterised by different values of flexible power associated to the appliances' cycles. Even though a great variability occurs, a common trend in the usage has been found early in the morning, at lunchtime and after dinner.

Based on collected data, Table 3 shows the archetypes energy characterisation as a result of the applied methodology. Such a characterisation has been used to preliminary estimate the flexible loads extent, according to scenario S0. The averaged values of Flexible Loads and Non-Flexible Loads are equal to 811 kWh/y and 1333 kWh/y, respectively.

**Table 3.** Archetypes reference parameters.

| Parameters | Archetype | | | | | | | | | | | | | |
|---|---|---|---|---|---|---|---|---|---|---|---|---|---|---|
| | #1 | #2 | #3 | #4 | #5 | #6 | #7 | #8 | #9 | #10 | #11 | #12 | #13 | #14 |
| $N_{Flex,S0}$ | 840 | 379 | 1030 | 172 | 1365 | 1261 | 1019 | 246 | 633 | 289 | 1868 | 650 | 984 | 504 |
| Flexible Loads ($N_{Flex,S0}$) [kWh/y] | 858 | 294 | 660 | 355 | 1758 | 927 | 661 | 188 | 1096 | 728 | 866 | 1366 | 957 | 637 |
| Non-Flexible Loads [kWh/y] | 2648 | 1024 | 1085 | 879 | 1298 | 1000 | 1099 | 881 | 2384 | 1218 | 1049 | 1754 | 1439 | 959 |

### 3.2. Flexible Loads by Strategy & Scenario: Real Flexibility

The simulated scenarios are characterised by the management strategy application together with different constraints setting up (see Table 2); by adding those technical limitations, the dwellings' capacity to participate in flexibility mechanisms is lessened.

The RF indicator use allows the authors to properly examine each scenario proposed by the authors. Indeed, once the RF variations associated to the archetypes over different months are known, it is possible to provide a realistic overview of potential flexibility. Table 4 summarises $N_{Flex,Sx}$ annual values for the selected archetypes, when the proposed scenarios have been accounted for.

**Table 4.** Archetypes $N_{Flex,Sx}$ by applying the Load Shifting strategy.

| Parameters | Archetype | | | | | | | | | | | | | |
|---|---|---|---|---|---|---|---|---|---|---|---|---|---|---|
| | #1 | #2 | #3 | #4 | #5 | #6 | #7 | #8 | #9 | #10 | #11 | #12 | #13 | #14 |
| $N_{Flex,S0}$ | 840 | 379 | 1030 | 172 | 1365 | 1261 | 1019 | 246 | 633 | 289 | 1868 | 650 | 984 | 504 |
| $N_{Flex,S1}$ | 432 | 236 | 526 | 83 | 753 | 534 | 769 | 160 | 334 | 194 | 1218 | 494 | 581 | 315 |
| $N_{Flex,S2}$ | 430 | 236 | 520 | 83 | 740 | 425 | 765 | 160 | 334 | 194 | 1216 | 492 | 560 | 315 |
| $N_{Flex,S3}$ | 339 | 236 | 201 | 83 | 661 | 386 | 684 | 160 | 265 | 167 | 1216 | 463 | 466 | 266 |
| $N_{Flex,S4}$ | 304 | 222 | 195 | 81 | 609 | 322 | 636 | 138 | 257 | 144 | 1037 | 414 | 367 | 259 |

According to what is reported in Table 4, the annual RF values vs. archetypes, with changes in scenarios, were calculated and are plotted in Figure 5. From data, it emerges how the more restrictive constraints applied in S2, S3 and S4 scenarios, lead to significant reductions of RF values.

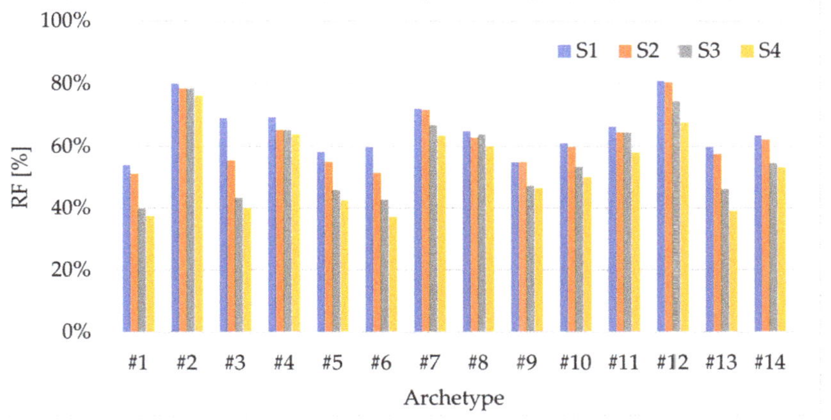


**Figure 5.** RF: archetype annual average values.

Specifically, RF's decrease from an average value of 66% once no limitation is applied in scenario (S1), to 53% taking into account all constraints in S4; moreover, a strong RF reduction has been registered between scenario S2 (RF = 62%) and S3 (RF = 56%), since limiting the energy-intensive appliances (i.e., washing machine, dish washer, air conditioner) hinders an effective load shifting of their cycles. Furthermore, that reduction is higher for those archetypes characterised by low RF values (e.g., #1, #3, #5, #6, #13). It is due to the fact that they largely use such appliances, and they are more affected by the constraint's introduction than the other archetypes.

Table 5 reports the $N_{Flex,Sx}$ monthly values in each scenario, having considered the 14 archetypes combination (i.e., summing all their contributions).

**Table 5.** Monthly $N_{Flex,Sx}$ by applying the Load Shifting strategy.

| Parameters | Months | | | | | | | | | | | |
|---|---|---|---|---|---|---|---|---|---|---|---|---|
| | Jan | Feb | Mar | Apr | May | Jun | Jul | Aug | Sept | Oct | Nov | Dec |
| $N_{Flex,S0}$ | 1878 | 1065 | 872 | 1235 | 924 | 640 | 792 | 597 | 1088 | 547 | 680 | 934 |
| $N_{Flex,S1}$ | 1220 | 693 | 589 | 795 | 603 | 404 | 301 | 234 | 561 | 356 | 425 | 515 |
| $N_{Flex,S2}$ | 1205 | 693 | 586 | 777 | 601 | 395 | 282 | 188 | 501 | 354 | 423 | 515 |
| $N_{Flex,S3}$ | 906 | 634 | 510 | 619 | 514 | 358 | 262 | 160 | 470 | 329 | 369 | 456 |
| $N_{Flex,S4}$ | 677 | 578 | 464 | 510 | 467 | 325 | 250 | 158 | 433 | 295 | 326 | 409 |

Figure 6 depicts the monthly RF values for each scenario, when archetypes have been gathered. Here too, RF reductions have been registered owing to the increasing limitations set up, according to S2, S3 and S4. Additionally, months belonging to the summertime are usually characterised by lower values than the others.

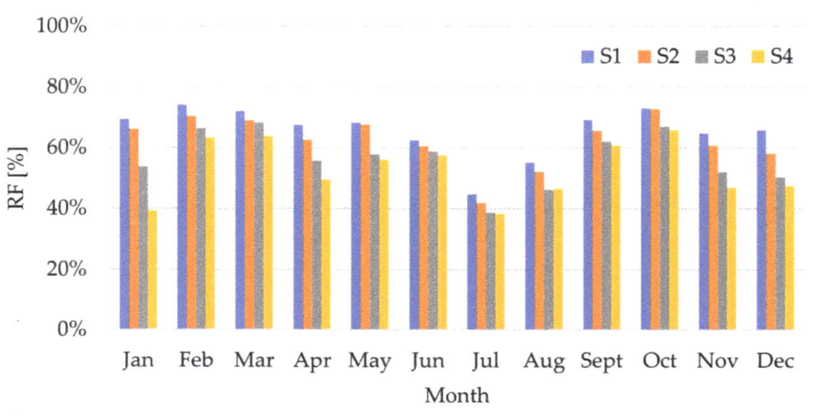


**Figure 6.** RF: monthly average values.

These RF reductions over the summertime depend mostly on the cooling systems operation, since the load shifting of air conditioner cycles is generally not feasible, due to the lack of suitable Load Reduction Time.

Thus, Figure 6 shows how, in June, July and August, the RF values are less sensitive to the introduction of S4 constraints. That behaviour is caused by climatic and time-delay factors, hailing from the larger photovoltaic power release in the national grid, and from lighting loads postponement. Moreover, the great number of Load Increase Time periods occurring in 10:00 a.m. up to 4:00 p.m., shrinks the loads shifting needs towards night-time.

More generally, where the Load Reduction Time periods are frequent, the RF is strongly penalised by applying the S4 features. Indeed, needs of loads shifting towards night-time are greater, but S4 constraints do not allow that.

### *3.3. Flexible Loads by Strategy & Scenario: Energy Shift*

It is important to highlight that RF is a relative value accounting only for appliance cycles. Therefore, it is not an exhaustive indicator for directly comparing each other the selected archetypes. For that purpose, a thorough analysis on loads peculiarities must be carried out. In such a way, it is possible to identify how often, the archetypes showing low RF, can provide higher flexibility in terms of energy shift (ES). For instance, referring to Figure 7, the archetypes number #6 and #13 are characterised by quite low ES values.

In the same way as before, for the RF indicator, Figure 8 depicts the ES average monthly values, when the archetypes have been clustered and constraint scenarios changed. That chart points out remarkable reductions in ES values over the summertime, owing to different operating mode of flexible appliances.

The archetypes' ES scattering values, sorted by months for a fixed scenario, are reported in Figures 9–12.

Comparing those charts, it is possible to recognise a progressive lowering in ES statistical distribution values over the year, as more technical limitations are added. In detail that reduction is greater where the ES values are higher, since shifted loads are generally much more, and hence, from a statistical point of view, even the unshifted loads are higher. Additionally, that issue repeats often in the winter months, where the ES third quartile amount to 26 kWh/month/dwelling, and peaks can get to 43 kWh/month/dwelling. By changing scenarios, those values go down until −32.5%, starting from an annual average of 27 kWh/month/dwelling up to 18 kWh/month/dwelling. Indeed, a high decrease is registered in wintertime, especially between the S1 and S4 scenarios. In that case, the average monthly values associated to January lessen from 40 to 17 kWh/month/dwelling. Differently, the reduction in the summertime is smaller, because of the modest amount of available flexibility, such as in June, where ES decreases from 19 to 14 kWh/month/dwelling.

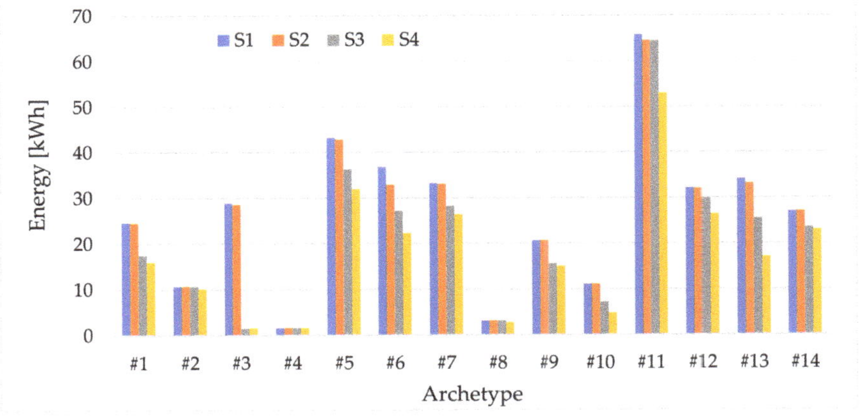


**Figure 7.** Archetypes average annual values of Monthly Energy Shift, with changes in constraint scenarios.

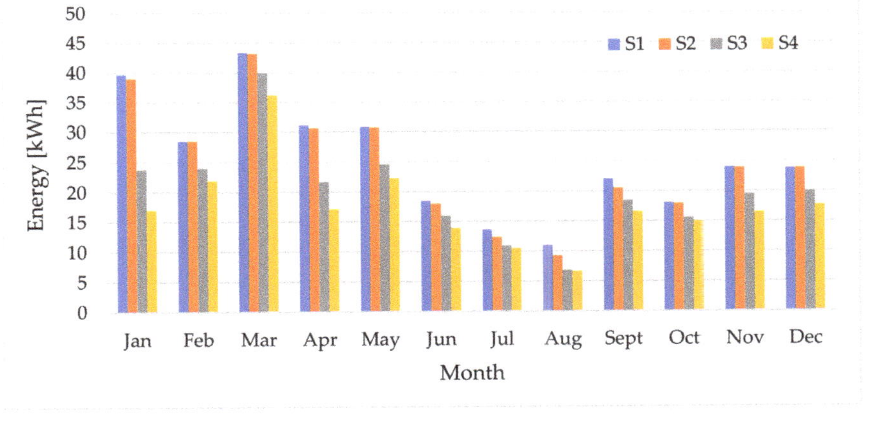


**Figure 8.** Clustered archetypes average Monthly Energy Shift, with changes in constraint scenarios.

The average values of monthly reduction, caused by applying technical constraints C1, C2 and C3 for simulating the S2 scenario, are not significant; conversely, the extreme limits of scattering values must be considered relevant (see Figure 10).

Introducing all of the restrictions associated to scenario S3, which mostly influence the energy-intensive appliances, the ES reductions become more evident. That feature especially occurs over the wintertime, where the ES values are greater, as reported in Figure 11. On the other hand, the average monthly variability values are mitigated, showing lower fluctuations over the whole year.

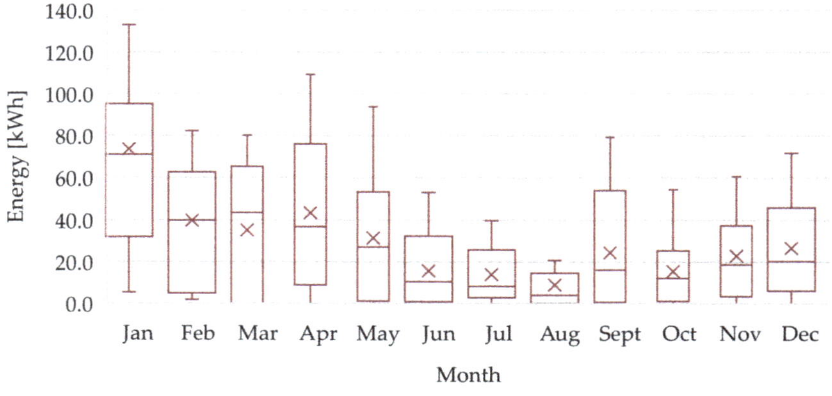


**Figure 9.** ES: Scenario 1, monthly average values.

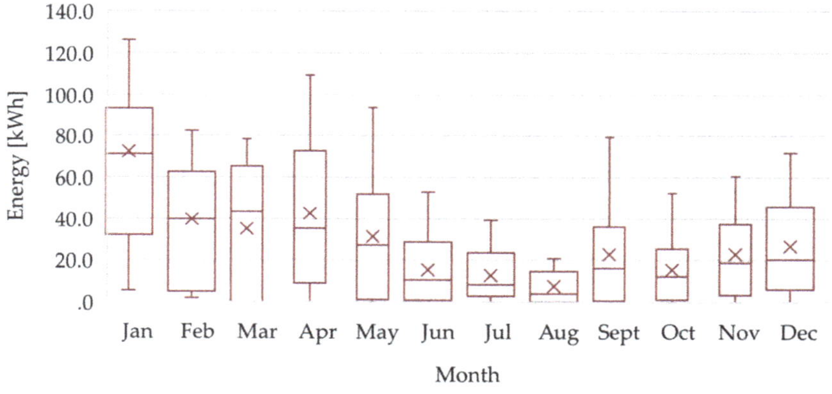


**Figure 10.** ES: Scenario 2, monthly average values.

Thereafter, when the most restrictive limitations are introduced by the S4 scenario, the ES monthly variability, in terms of both average and extreme scattering values, is characterised by a further lessening along with a more flattened trend. Such a behaviour, it is due to the fact that, the current technical constraints deny loads shifting through the night time (see Figure 12).

### 3.4. *Flexible Loads by Strategy & Scenario: Peak Shaving*

The Peak Shaving indicator trends (i.e., PS) have been clearly plotted in Figures 13 and 14. It is worth noticing that PS is defined as the maximum registered power reduction, over a 15-min time span. That indicator is calculated for each archetype as well as for the clustered version. In so doing, it is possible to evaluate firstly the flexibility system potential, and secondly, to identify the number of minimum users able to provide the grid DSO with specific DSM parameters and references. Moreover, from Figure 13, it is possible to distinguish those archetypes using the most energy intensive appliances, so that greater DSM services might be potentially provided to the grid. In Figure 14, the monthly variation shows a modest reduction only in the summertime. The PS variability between the simulated constraint scenarios remains almost constant, highlighting the nature of this indicator. Indeed, it indicates a maximum value instead of an average one, as for the previous indicators. In detail, PS indicator small and quite negligible reductions are detected by applying the different scenarios, namely: 1678 W for S1, 1677 W for S2, 1612 W for S3, 1592 W for S4.

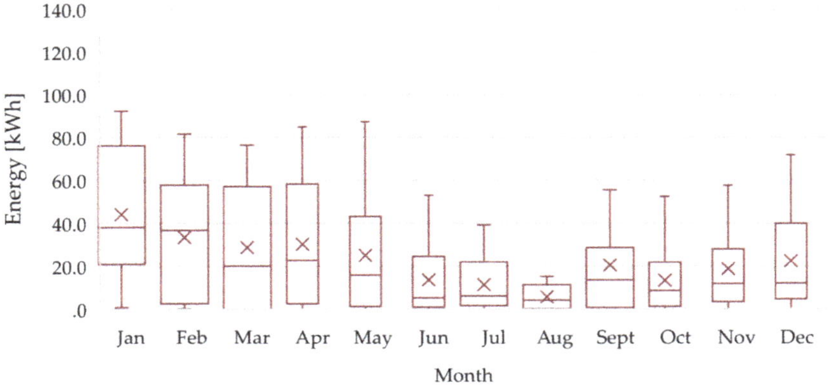


**Figure 11.** ES: Scenario 3, monthly average values.

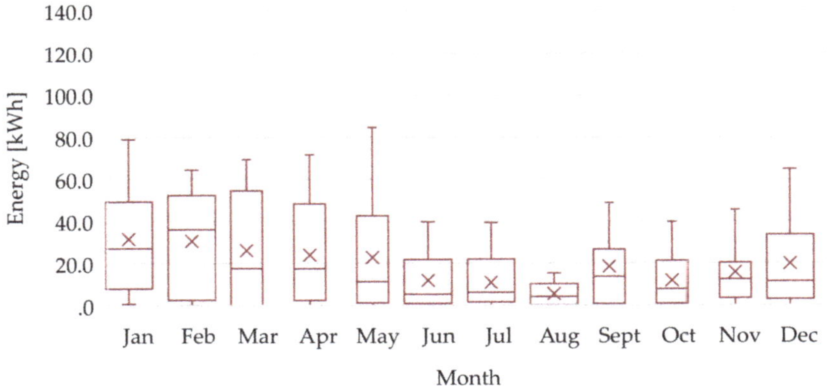


**Figure 12.** ES: Scenario 4, monthly average values.

The PS scattered values associated to the archetypes, sorted by months, are plotted in Figures 15–18. Those charts represent the maximum shaved power distribution, caused by the scenario Sx application. From data analysis, it is possible to conclude that the higher variations correspond to the lower PS values. That feature entails that some archetypes keep constant their PS values over the year, while the others are characterised by lower PS, generally over the summertime. Moreover, in Figure 15, it can be noticed how the median monthly values do not exceed the threshold of 2000 W, which mostly range between 1500 W and 1900 W.

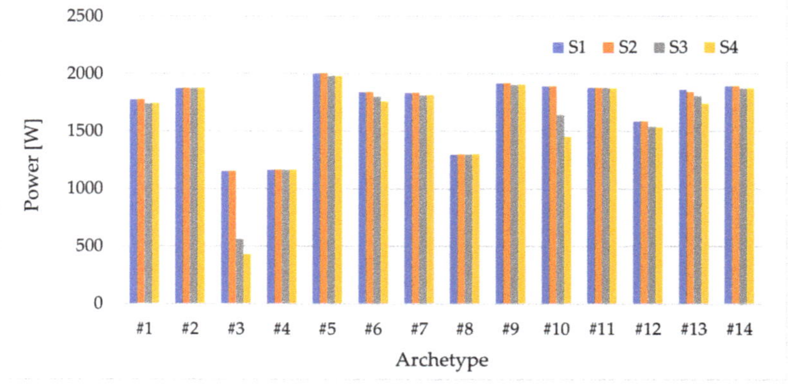


**Figure 13.** PS: archetype annual average values.

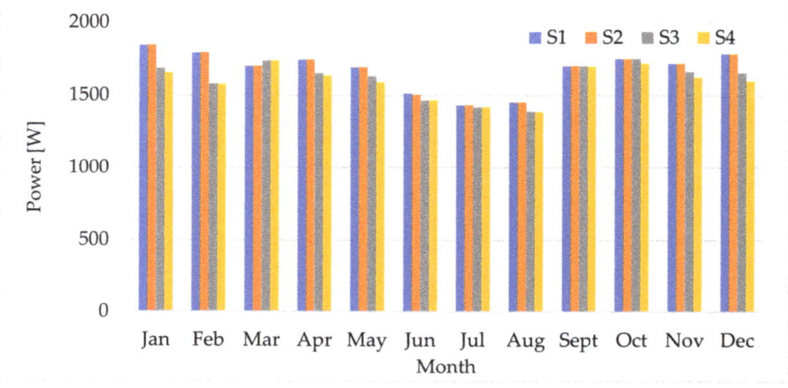


**Figure 14.** PS: monthly average values.

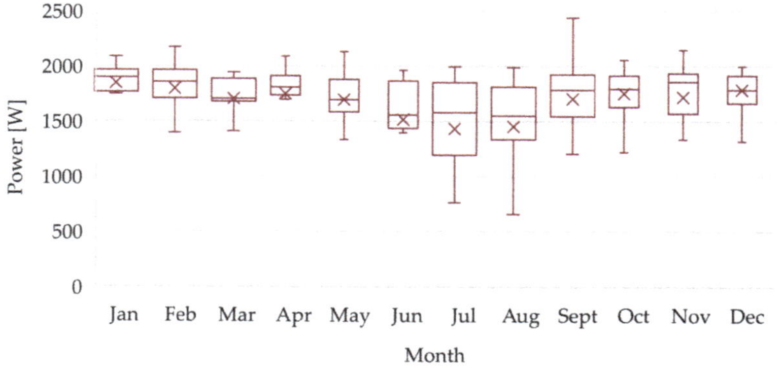


**Figure 15.** PS: Scenario 1, monthly average values.

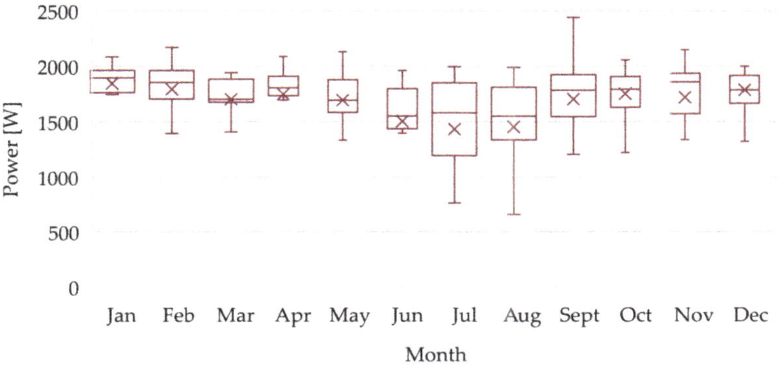


**Figure 16.** PS: Scenario 2, monthly average values.

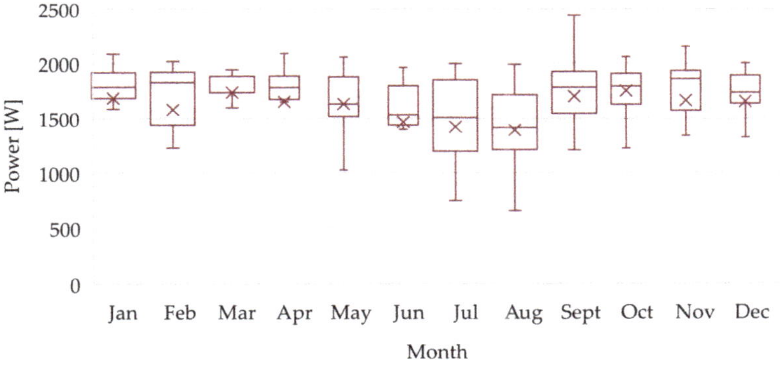


**Figure 17.** PS: Scenario 3, monthly average values.

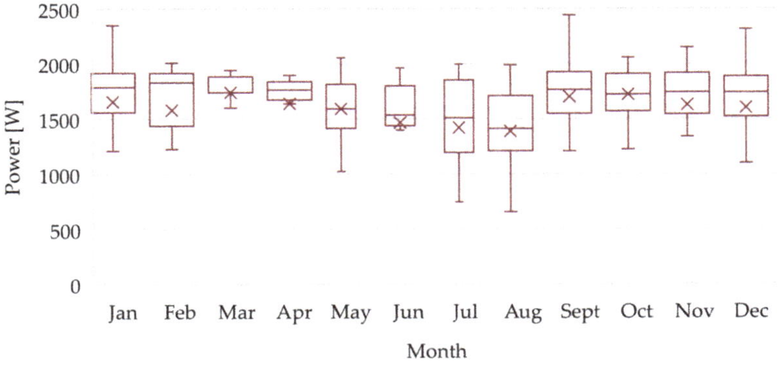


**Figure 18.** PS: Scenario 4, monthly average values.

Thus, Figure 16 reports the PS variation owing to the S2 scenario application. Comparing these data with those related to the S1 scenario, small differences are registered, exception for the minimum values of the monthly scattering plot.

That issue is emphasised further within the chart associated to the S3 scenario (see Figure 17), where the minimum values further decrease (first quartile), hence the PS variation range increases (e.g., in February).

The same trend is confirmed by the S4 scenario (see Figure 18), where a modest lessening in the maximum values, together with the median and the average, occur. The average annual values drift away, starting from 1777 W for the S1 scenario, towards 1600 W for the S4 scenario.

## 4. Conclusions

This work aims at defining and quantifying the real flexibility offered by a building cluster, in the Italian residential sector, to the national grid. In so doing, that cluster can participate in a demand response program in accordance with a Flexibility Price-market-based Strategy. Extrapolating and exploiting data collection, hailing from an experimental campaign carried out on 14 reference dwellings, over two years, a Load Shifting Strategy has been applied. Additionally, some theoretical indicators have been presented and extensively discussed on the basis of a real case study. Then, four different scenarios have been built by imposing several technical constraints on the household appliances, accounting for operation contemporaneity, building occupancy, cycles time sequences etc. All of scenarios are characterised by a growing limitation degree in the Load Shifting strategy, which consequently causes considerable reduction of Real Flexibility capacity that a building cluster can provide.

In order to evaluate the potential reduction hailing from the realistic constraints, a reference scenario S0 has been built. It represents the theoretical Flexibility potential. In detail, scenario S1 includes only the Load Shifting Strategy application; in scenarios, S2–S4 seven technical constraints have been added. Starting from the RF average value over the year equal to 66%, it decreases up to 53%, with seasonal differences which get the lowest values in July and August. RF values up to 80%, can be provided by some sample dwellings. Notwithstanding, when buildings are clustered, the virtuous end users' habits can be mitigated, leading to lower flexibility capacity associated to the residential sector.

Other indicators have been calculated, and one of these is the ES (energy amount effectively moved over the day deriving from the Load Shifting strategy). Monthly average values of ES decrease from 12 to 8 kWh/month/dwelling, due to the more restrictive scenario application (i.e., S4). ES peak values have been calculated and 35 kWh/month/dwelling has been accomplished. The PS indicator has been calculated too: values close to 1600 W, with a low sensitivity to the different scenarios, show moderate effects in terms of power reduction; nevertheless, by gathering some dwellings, it is possible to get to higher PSs for providing the grid with an efficient strategy to reshape the load curve.

The main outcomes from simulation can be outlined as follows:

- The RF value shrinks as the constraints number increases; starting from S1 to S4 the registered RFs are equal to 66%; 62%; 56%; 53%, respectively.
- The ES value decreases by changing scenarios; from S1 to S4, 27; 26; 21; 18 kWh/month/dwelling, have been registered, respectively.
- The value of PS does not significantly decrease, from S1 to S4, e.g., 1678 W; 1677 W; 1612 W; 1592 W, respectively.
- The appliances' cycles are mostly shifted in the afternoon, between 4:00 p.m. and 8:00 p.m.; the time span in which they are moved is close to the late evening and night-time. As regards the S4 scenario, the loads shifting occurs early in the morning (after 6:00 a.m.).
- The Dish Washer and Washing Machine cycles that generally are shifted, have an energy consumption approximately equal to 1 kWh/cycle.

- In the summertime (from June to September) flexibility is on the average lower than in the other months, in terms of both RF and ES, because of the different composition of summer loads (e.g., lighting, air conditioning).
- The indicators values reduction is stronger in S2 and S3 rather than in the other scenarios. That is due to the restrictive limitations addition which have been applied to the energy-intensive appliances.

This article is presented based on the actual data of 14 homes, which are characterised by a low electrification degree; among those, only one reference home is equipped with a photovoltaic plant. New scenarios are being developed dealing with dwellings which use heat pumps for heating and DHW purpose, and which have self-generation hailing from PV along with batteries pack. In these cases, self-generation and storage devices constitute further constraints compared to those examined in this article. As a result, the flexibility potential that each individual dwelling can offer to the electric spot market will be reduced.

**Author Contributions:** F.M. contributed to this paper by the conceptualization and validation; G.L.B. provided the formal analysis and was the writer of the final draft along with the reviewing and editing process; J.C. was the responsible for data collection, their postprocessing and software implementation; F.M. and S.R. were the project scientific coordinators: the former was responsible of funding acquisition along with project administration; the latter was the general project supervisor. All authors have read and agreed to the published version of the manuscript.

**Funding:** This research received no external funding.

**Acknowledgments:** This work is a part of a wider research activity dealing with: "Technologies for efficient use and deployment of electric energy carrier". The project has been carried out in cooperation with ENEA-DTE (Italian National Agency for New Technologies, Energy and Sustainable Economic Development-Department of Energy Technologies) and CITERA (Sapienza University of Rome-Interdepartmental Research Centre for Territory, Construction, Restoration and Environment). The aforementioned institutions are gratefully acknowledged by the authors for their support and funding.

**Conflicts of Interest:** The authors declare no conflict of interest.

## Appendix A

**Table A1.** Strategy to optimise the load shifting: weekdays; (−2, Green) Strong Load reduction; (−1, Light Green) Weak Load re-duction; (0, White) No Load variation; (1, Light Red) Weak Load increase; (2, Red) Strong Load increase.

| Months | Hours | | | | | | | | | | | | | | | | | | | | | | | |
|---|---|---|---|---|---|---|---|---|---|---|---|---|---|---|---|---|---|---|---|---|---|---|---|---|
| | **1** | **2** | **3** | **4** | **5** | **6** | **7** | **8** | **9** | **10** | **11** | **12** | **13** | **14** | **15** | **16** | **17** | **18** | **19** | **20** | **21** | **22** | **23** | **24** |
| January | 1 | 2 | 2 | 2 | 2 | 2 | 1 | −1 | −2 | 0 | 0 | 0 | 1 | 0 | 0 | −1 | 0 | −1 | −1 | −2 | −1 | −1 | 0 | 0 |
| February | 2 | 2 | 2 | 2 | 2 | 2 | −1 | −1 | −1 | 0 | 0 | 1 | 1 | 1 | 1 | −1 | −1 | −1 | −1 | −2 | −2 | −1 | 0 | 0 |
| March | 1 | 2 | 2 | 2 | 2 | 2 | −1 | −1 | −1 | 0 | 0 | 1 | 1 | 1 | 0 | −1 | −1 | 0 | −1 | −2 | −2 | −1 | 0 | 0 |
| April | 1 | 2 | 2 | 2 | 2 | 2 | 1 | −1 | −1 | 0 | 0 | 0 | 1 | 1 | 0 | 0 | −1 | −1 | −1 | −2 | −2 | −2 | −1 | 0 |
| May | 1 | 2 | 2 | 2 | 2 | 2 | 1 | −2 | −1 | 0 | 0 | 0 | 1 | 0 | 0 | 0 | −1 | 0 | −1 | −2 | −2 | −2 | −1 | 0 |
| June | 1 | 2 | 2 | 2 | 2 | 2 | 1 | −1 | 0 | 0 | 0 | 1 | 1 | 0 | 0 | −1 | −1 | −1 | −1 | −2 | −2 | −2 | −1 | 0 |
| July | 0 | 1 | 2 | 2 | 2 | 2 | 2 | 1 | 0 | 0 | 0 | 0 | 1 | 0 | 0 | −1 | −1 | −1 | −2 | −2 | −2 | −2 | −1 | 0 |
| August | −1 | 1 | 2 | 2 | 2 | 2 | 2 | 1 | 0 | 0 | 1 | 1 | 1 | 0 | 0 | −1 | −1 | −2 | −1 | −1 | −2 | −2 | −1 | −1 |
| September | 1 | 2 | 2 | 2 | 2 | 2 | 1 | −1 | 0 | −1 | 0 | 0 | 1 | 1 | 0 | −1 | −1 | −1 | −1 | −2 | −2 | −1 | 0 | 0 |
| October | 2 | 2 | 2 | 2 | 2 | 2 | 1 | −1 | −1 | −1 | 0 | 0 | 1 | 1 | 1 | −1 | −1 | −1 | −1 | −2 | −2 | −1 | 0 | 0 |
| November | 2 | 2 | 2 | 2 | 2 | 2 | 1 | −1 | −1 | 0 | 0 | −1 | 1 | 1 | 0 | −1 | −1 | −1 | −1 | −2 | −1 | 0 | 0 | 0 |
| December | 2 | 2 | 2 | 2 | 2 | 2 | 1 | −1 | 0 | −1 | 0 | 0 | 1 | 1 | −1 | −1 | −1 | −2 | −2 | −2 | −1 | 0 | 0 | 0 |

**Table A2.** Strategy to optimise the load shifting: Saturdays. (−2, Green) Strong Load reduction; (−1, Light Green) Weak Load reduction; (0, White) No Load variation; (1, Light Red) Weak Load increase; (2, Red) Strong Load increase.

| Months | Hours | | | | | | | | | | | | | | | | | | | | | | | |
|---|---|---|---|---|---|---|---|---|---|---|---|---|---|---|---|---|---|---|---|---|---|---|---|---|
| | 1 | 2 | 3 | 4 | 5 | 6 | 7 | 8 | 9 | 10 | 11 | 12 | 13 | 14 | 15 | 16 | 17 | 18 | 19 | 20 | 21 | 22 | 23 | 24 |
| January | 1 | 2 | 2 | 2 | 2 | 2 | 1 | 0 | 0 | −1 | −1 | 0 | 1 | 0 | 0 | 0 | −1 | −1 | −2 | −2 | −2 | −1 | −1 | 1 |
| February | 1 | 1 | 2 | 2 | 2 | 2 | 1 | 0 | 0 | −1 | −1 | 0 | 1 | 0 | 0 | 0 | −1 | −1 | −1 | −1 | −2 | −1 | −1 | 1 |
| March | 0 | 1 | 1 | 2 | 2 | 1 | 0 | 0 | 0 | −1 | −1 | 1 | 0 | 0 | 0 | 0 | 0 | 0 | 0 | −2 | −2 | −2 | −1 | 0 |
| April | 0 | 1 | 1 | 1 | 2 | 1 | 0 | 0 | 0 | −2 | −1 | 1 | 0 | 0 | 0 | 0 | 0 | 0 | 0 | −2 | −2 | −2 | −1 | 0 |
| May | 0 | 0 | 1 | 2 | 2 | 2 | 1 | 0 | 0 | −1 | 0 | 0 | 0 | 0 | 0 | 0 | 0 | 0 | −1 | −1 | −2 | −2 | −1 | 1 |
| June | 0 | 0 | 0 | 1 | 1 | 1 | 1 | 1 | 0 | −1 | 0 | 1 | 0 | 0 | 0 | 0 | 0 | 0 | −1 | −2 | −2 | −2 | −2 | 0 |
| July | 0 | 0 | 0 | 0 | 1 | 1 | 1 | 1 | 0 | −1 | 0 | 0 | 0 | 0 | 0 | 0 | 0 | 0 | −1 | −1 | 0 | 0 | 0 | −1 |
| August | −1 | 0 | 0 | 0 | 1 | 1 | 1 | 1 | 0 | 1 | 1 | 0 | 1 | 0 | 0 | 0 | 0 | −1 | 0 | −1 | −2 | −1 | −2 | −1 |
| September | 0 | 0 | 1 | 1 | 1 | 1 | 0 | 0 | 0 | 0 | 0 | 0 | 0 | 0 | 0 | 1 | 0 | −1 | −1 | −2 | −2 | −2 | −1 | 1 |
| October | 0 | 0 | 1 | 1 | 2 | 1 | 0 | 0 | −1 | −1 | 0 | 1 | 0 | 0 | 0 | 1 | 0 | −1 | 0 | −2 | −2 | −1 | 0 | 0 |
| November | 1 | 1 | 1 | 2 | 2 | 2 | 1 | 0 | 0 | −1 | −1 | 0 | 0 | 0 | 0 | 0 | −2 | −2 | −1 | −1 | −2 | −1 | 0 | 1 |
| December | 1 | 1 | 2 | 2 | 2 | 2 | 2 | 0 | 0 | −1 | −1 | 0 | 1 | 1 | 0 | −1 | −2 | −2 | −1 | −2 | −2 | −1 | 0 | 0 |

**Table A3.** Strategy to optimise the load shifting: non-working days. (−2, Green) Strong Load reduction; (−1, Light Green) Weak Load reduction; (0, White) No Load variation; (1, Light Red) Weak Load increase; (2, Red) Strong Load increase.

| Months | Hours | | | | | | | | | | | | | | | | | | | | | | | |
|---|---|---|---|---|---|---|---|---|---|---|---|---|---|---|---|---|---|---|---|---|---|---|---|---|
| | 1 | 2 | 3 | 4 | 5 | 6 | 7 | 8 | 9 | 10 | 11 | 12 | 13 | 14 | 15 | 16 | 17 | 18 | 19 | 20 | 21 | 22 | 23 | 24 |
| January | 0 | 1 | 2 | 2 | 2 | 2 | 2 | 2 | 1 | 1 | −1 | −1 | −1 | 0 | 0 | 0 | −2 | −1 | −1 | −2 | −2 | −2 | −1 | 1 |
| February | 0 | 1 | 2 | 2 | 2 | 2 | 1 | 1 | 0 | 0 | −1 | 0 | 0 | 0 | 0 | 0 | −1 | 0 | −1 | −2 | −2 | −2 | −2 | 0 |
| March | −1 | 1 | 2 | 2 | 2 | 2 | 1 | 1 | 1 | −1 | −1 | −1 | 0 | 1 | 0 | 0 | −1 | 0 | 0 | −2 | −2 | −1 | −1 | 0 |
| April | 0 | 1 | 1 | 1 | 1 | 0 | 0 | 0 | 0 | −1 | 0 | 0 | 0 | 1 | 1 | 1 | 0 | 0 | −1 | −2 | −2 | −2 | −2 | −1 |
| May | 0 | 0 | 1 | 2 | 1 | 0 | 1 | 1 | 0 | 0 | −1 | 0 | 0 | 0 | 0 | 0 | 0 | 1 | 0 | −1 | −2 | −2 | −2 | −1 |
| June | 0 | 0 | 0 | 1 | 1 | 1 | 2 | 1 | 1 | 1 | −1 | 1 | 1 | 0 | 0 | 0 | −1 | −1 | −1 | −2 | −2 | −2 | −2 | −1 |
| July | −1 | 0 | 0 | 0 | 0 | 1 | 2 | 1 | 1 | 0 | 1 | 1 | 1 | 0 | 0 | 0 | 0 | −1 | −1 | −1 | −2 | −1 | −1 | 0 |
| August | 0 | 0 | 0 | 0 | 1 | 0 | 1 | 1 | 1 | 1 | 1 | 0 | 1 | 0 | 0 | 0 | 0 | −1 | −1 | −1 | −1 | −2 | −1 | −1 |
| September | 0 | 0 | 1 | 2 | 2 | 2 | 0 | 1 | 1 | −1 | 0 | 1 | 0 | 0 | 0 | 0 | 0 | −1 | −2 | −2 | −2 | −2 | −1 | 0 |
| October | 0 | 1 | 1 | 2 | 2 | 2 | 1 | 1 | 0 | −1 | −1 | 0 | 0 | 0 | 0 | 0 | 0 | 0 | −1 | −2 | −2 | −2 | −1 | 0 |
| November | 1 | 2 | 2 | 2 | 2 | 2 | 2 | 1 | 1 | 1 | −1 | 0 | 0 | 0 | −1 | −1 | −1 | −1 | −2 | −2 | −2 | −2 | −1 | 0 |
| December | 1 | 1 | 2 | 2 | 2 | 2 | 2 | 1 | 1 | −1 | −1 | −1 | −1 | 1 | 1 | −1 | −2 | −1 | −2 | −2 | −2 | −2 | −1 | 1 |

## Appendix B

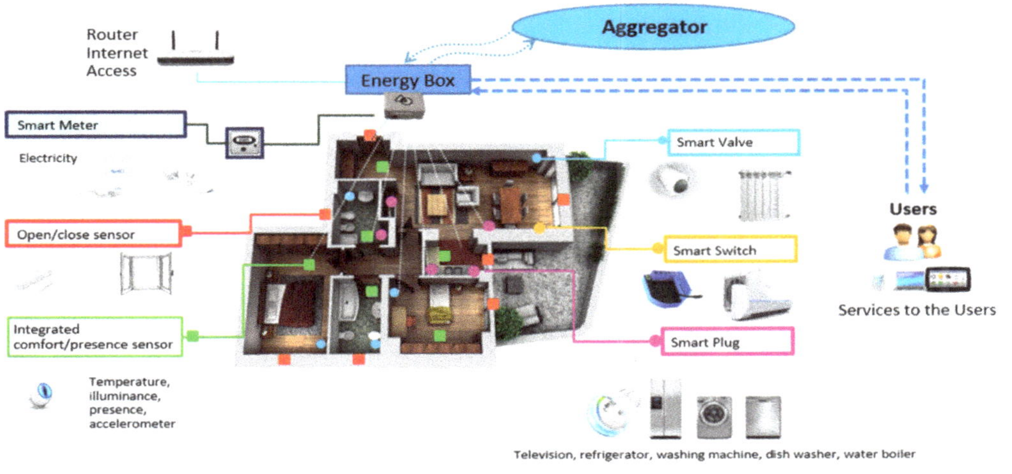


**Figure A1.** Control kit layout.

**Table A4.** Control kits configuration for the archetypes.

| Function | Device | Archetype | | | | | | | | | | | | | |
|---|---|---|---|---|---|---|---|---|---|---|---|---|---|---|---|
| | | #1 | #2 | #3 | #4 | #5 | #6 | #7 | #8 | #9 | #10 | #11 | #12 | #13 | #14 |
| Energy box | Gateway | 1 | 1 | 1 | 1 | 1 | 1 | 1 | 1 | 1 | 1 | 1 | 1 | 1 | 1 |
| Monitoring | Electricity meters | 1 | 1 | 1 | 1 | 2 | 1 | 1 | 1 | 1 | 1 | 1 | 1 | 1 | 2 |
| | Multi-sensors (temperature, presence, brightness) | 5 | 6 | 6 | 4 | 6 | 6 | 4 | 4 | 7 | 6 | 3 | 9 | 7 | 7 |
| | Windows/doors opening and closing detectors | 7 | 8 | 6 | 5 | 8 | 8 | 5 | 5 | 10 | 10 | 6 | 9 | 12 | 9 |
| Control | Smart Valves | 6 | 5 | 0 | 4 | 3 | 6 | 5 | 3 | 8 | 6 | 0 | 0 | 7 | 0 |
| | Smart Plugs | 4 | 3 | 4 | 4 | 3 | 4 | 4 | 3 | 3 | 4 | 3 | 5 | 3 | 6 |
| | Smart Switches | 1 | 0 | 0 | 0 | 1 | 1 | 1 | 1 | 1 | 1 | 0 | 1 | 0 | 0 |

**Table A5.** Archetypes appliances and characteristics.

| Archetype | Floor Surface [m$^2$] | Heating and DHW * | Cooling * | PV Array | WM ** | DW ** | TD ** |
|---|---|---|---|---|---|---|---|
| #1 | 49 | NCB | 2 HP | | 7; 5; A+ | 6; 7; A | |
| #2 | 101 | NCB | 1 HP | | 10; 2.5; A | | |
| #3 | 100 | NCB | 1 HP | | 7; 5; A+ | | |
| #4 | 50 | NCB | 1 HP | | 7; 1.5; A+ | | 5; 0.5; A |
| #5 | 100 | CB + HP | 4 HP | | 7; 4; A++ | 5; 4; A | 5; 4; A |
| #6 | 65 | CB | 3 HP | | 7; 6; A | 12; 3.5; A | 7; 0.5; B |
| #7 | 65 | NCB | 1 HP | | 7; 5; A+ | 6; 7; A | |
| #8 | 60 | CB | | | 7; 2; A++ | 12; 1.5; A+ | |
| #9 | 95 | NCB | 2 HP | | 7; 5; A−++ | 12; 8; A+ | |

**Table A5.** *Cont.*

| Archetype | Floor Surface [m$^2$] | Heating and DHW * | Cooling * | PV Array | WM ** | DW ** | TD ** |
|---|---|---|---|---|---|---|---|
| #10 | 102 | NCB | 1 HP | | 7; 3; A+ | 14; 5; A | |
| #11 | 67 | CB | | | 10; 5; B | 6; 5; B | |
| #12 | 134 | CB | | | 7; 6; A | 14; 7; A | 6; 3; B |
| #13 | 124 | CB | | | 5; 4; A | 12; 7; A+ | |
| #14 | 123 | NCB + solar collectors | | 3.9 kW | 5; 4; A | 12; 7; A+ | |

* NCB: Non-Condensing Boiler; CB: Condensing Boiler; HP: Heat Pump; ** WM: Washing machine; DW: Dishwasher; TD: Tumble dryer; Capacity, cycles per week, Energy Class.

**Table A6.** Family composition of each archetype.

| Archetype | Occupants * | Description |
|---|---|---|
| #1 | 4; (1; 3; 4; 4) | Family with two teenage children and one unemployed parent |
| #2 | 2; (0; 0; 2; 2) | Commuter Workers |
| #3 | 4; (0; 3; 4; 4) | Family with school-aged children, and one part-time working parent |
| #4 | 1; (0; 0; 1; 1) | Commuter Worker |
| #5 | 4; (1; 3; 4; 4) | Family with school-aged children, and one home parent |
| #6 | 4; (1; 3; 4; 4) | Family with school-aged children and babies, and one unemployed parent |
| #7 | 3; (0; 0; 3; 3) | Family with a baby and commuter parents |
| #8 | 2; (1; 1; 2; 2) | Commuter worker, awaiting employment |
| #9 | 3; (1; 2; 3; 3) | Family with a school-aged child, and one commuter worker |
| #10 | 2; (0; 1; 2; 2) | Family of commuter workers |
| #11 | 3; (0; 2; 3; 3) | Family with a school-aged child, and two commuter workers |
| #12 | 4; (0; 1; 4; 4) | Family with two adult children, two commuter parents |
| #13 | 2; (0; 1; 2; 2) | Family with a school-aged child, and two commuter workers |
| #14 | 2; (2; 2; 2; 2) | Two Pensioners |

* Number of occupants; (8 a.m. to 1 p.m.; 1 p.m. to 7 p.m.; 7 p.m. to 12 a.m.; 12 a.m. to 8 a.m.).

## References


1. European Commission. *A Clean Planet for All A European Strategic Long-Term Vision for a Prosperous, Modern, Competitive and Climate Neutral Economy*; European Commission: Brussels, Belgium, 2018.
2. Brouwer, A.S.; Van Den Broek, M.; Seebregts, A.; Faaij, A. Impacts of large-scale Intermittent Renewable Energy Sources on electricity systems, and how these can be modeled. *Renew. Sustain. Energy Rev.* **2014**, *33*, 443–466. [CrossRef]
3. Schibuola, L.; Scarpa, M.; Tambani, C. Demand response management by means of heat pumps controlled via real time pricing. *Energy Build.* **2015**, *90*, 15–28. [CrossRef]
4. Haider, H.T.; See, O.H.; Elmenreich, W. A review of residential demand response of smart grid. *Renew. Sustain. Energy Rev.* **2016**, *59*, 166–178. [CrossRef]
5. Magnago, F.H.; Alemany, J.; Lin, J. Impact of demand response resources on unit commitment and dispatch in a day-ahead electricity market. *Int. J. Electr. Power Energy Syst.* **2015**, *68*, 142–149. [CrossRef]
6. Eurostat Statistics | Eurostat/. Available online: https://ec.europa.eu/eurostat/databrowser/view/ten00124/default/table?lang=en (accessed on 31 March 2020).
7. Foteinaki, K.; Li, R.; Heller, A.; Rode, C. Heating system energy flexibility of low-energy residential buildings. *Energy Build.* **2018**, *180*, 95–108. [CrossRef]
8. Jensen, S.Ø.; Marszal-Pomianowska, A.; Lollini, R.; Pasut, W.; Knotzer, A.; Engelmann, P.; Stafford, A.; Reynders, G. IEA EBC Annex 67 Energy Flexible Buildings. *Energy Build.* **2017**, *155*, 25–34. [CrossRef]
9. Rahmani-Andebili, M. Scheduling deferrable appliances and energy resources of a smart home applying multi-time scale stochastic model predictive control. *Sustain. Cities Soc.* **2017**, *32*, 338–347. [CrossRef]
10. Chen, Y.; Xu, P.; Gu, J.; Schmidt, F.; Li, W. Measures to improve energy demand flexibility in buildings for demand response (DR): A review. *Energy Build.* **2018**, *177*, 125–139. [CrossRef]

11. Cumo, F.; Curreli, F.R.; Pennacchia, E.; Piras, G.; Roversi, R. Enhancing the urban quality of life: A case study of a coastal city in the metropolitan area of Rome. *WIT Trans. Built Environ.* **2017**, *170*, 127–137. [CrossRef]
12. Péan, T.; Costa-Castelló, R.; Salom, J. Price and carbon-based energy flexibility of residential heating and cooling loads using model predictive control. *Sustain. Cities Soc.* **2019**, *50*, 101579. [CrossRef]
13. De Santoli, L.; Lo Basso, G.; Astiaso Garcia, D.; Piras, G.; Spiridigliozzi, G. Dynamic Simulation Model of Trans-Critical Carbon Dioxide Heat Pump Application for Boosting Low Temperature Distribution Networks in Dwellings. *Energies* **2019**, *12*, 484. [CrossRef]
14. Mazzoni, S.; Ooi, S.; Nastasi, B.; Romagnoli, A. Energy storage technologies as techno-economic parameters for master-planning and optimal dispatch in smart multi energy systems. *Appl. Energy* **2019**, *254*, 113682. [CrossRef]
15. Nastasi, B. Hydrogen policy, market, and R & D projects. In *Solar Hydrogen Production: Processes, Systems and Technologies*; Elsevier: Amsterdam, The Netherlands, 2019; pp. 31–44. ISBN 9780128148549.
16. Nastasi, B.; Lo Basso, G.; Astiaso Garcia, D.; Cumo, F.; de Santoli, L. Power-to-gas leverage effect on power-to-heat application for urban renewable thermal energy systems. *Int. J. Hydrogen Energy* **2018**, *43*, 23076–23090. [CrossRef]
17. Roversi, R.; Cumo, F.; D'Angelo, A.; Pennacchia, E.; Piras, G. Feasibility of municipal waste reuse for building envelopes for near zero-energy buildings. *WIT Trans. Ecol. Environ.* **2017**, *224*, 115–125. [CrossRef]
18. D'Ettorre, F.; De Rosa, M.; Conti, P.; Testi, D.; Finn, D. Mapping the energy flexibility potential of single buildings equipped with optimally-controlled heat pump, gas boilers and thermal storage. *Sustain. Cities Soc.* **2019**, *50*. [CrossRef]
19. Mancini, F.; Romano, S.; Lo Basso, G.; Cimaglia, J.; De Santoli, L. How the italian residential sector could contribute to load flexibility in demand response activities: A methodology for residential clustering and developing a flexibility strategy. *Energies* **2020**, *13*, 3359. [CrossRef]
20. Lucas, A.; Jansen, L.; Andreadou, N.; Kotsakis, E.; Masera, M. Load Flexibility Forecast for DR Using Non-Intrusive Load Monitoring in the Residential Sector. *Energies* **2019**, *12*, 2725. [CrossRef]
21. Mancini, F.; Lo Basso, G.; De Santoli, L. Energy Use in Residential Buildings: Characterisation for Identifying Flexible Loads by Means of a Questionnaire Survey. *Energies* **2019**, *12*, 2055. [CrossRef]
22. Siano, P.; Sarno, D. Assessing the benefits of residential demand response in a real time distribution energy market. *Appl. Energy* **2016**, *161*, 533–551. [CrossRef]
23. Agbonaye, O.; Keatley, P.; Huang, Y.; Bani-mustafa, M.; Hewitt, N. Design, Valuation and Comparison of Demand Response Strategies for Congestion Management. *Energies* **2020**, *13*, 6085. [CrossRef]
24. Rashid, M.M.U.; Granelli, F.; Hossain, M.A.; Alam, M.S.; Al-Ismail, F.S.; Karmaker, A.K.; Rahaman, M.M. Development of Home Energy Management Scheme for a Smart Grid Community. *Energies* **2020**, *13*, 4288. [CrossRef]
25. Yao, L.; Hashim, F.H.; Lai, C.-C. Dynamic Residential Energy Management for Real-Time Pricing. *Energies* **2020**, *13*, 2562. [CrossRef]
26. Mancini, F.; Basso, G.L.; Santoli, L.D. Energy use in residential buildings: Impact of building automation control systems on energy performance and flexibility. *Energies* **2019**, *12*, 2896. [CrossRef]
27. Afzalan, M.; Jazizadeh, F. Residential loads flexibility potential for demand response using energy consumption patterns and user segments. *Appl. Energy* **2019**, *254*, 113693. [CrossRef]
28. D'hulst, R.; Labeeuw, W.; Beusen, B.; Claessens, S.; Deconinck, G.; Vanthournout, K. Demand response flexibility and flexibility potential of residential smart appliances: Experiences from large pilot test in Belgium. *Appl. Energy* **2015**, *155*, 79–90. [CrossRef]
29. Finck, C.; Li, R.; Zeiler, W. Optimal control of demand flexibility under real-time pricing for heating systems in buildings: A real-life demonstration. *Appl. Energy* **2020**, *263*, 114671. [CrossRef]
30. Feuerriegel, S.; Neumann, D. Integration scenarios of Demand Response into electricity markets: Load shifting, financial savings and policy implications. *Energy Policy* **2016**, *96*, 231–240. [CrossRef]
31. Shirazi, E.; Jadid, S. Cost reduction and peak shaving through domestic load shifting and DERs. *Energy* **2017**, *124*, 146–159. [CrossRef]
32. Zheng, M.; Meinrenken, C.J.; Lackner, K.S. Smart households: Dispatch strategies and economic analysis of distributed energy storage for residential peak shaving. *Appl. Energy* **2015**, *147*, 246–257. [CrossRef]

energies

Article

# Optimizing Power and Heat Sector Coupling for the Implementation of Carbon-Free Communities

Arslan Ahmad Bashir [1,*], Andreas Lund [1,2], Mahdi Pourakbari-Kasmaei [1] and Matti Lehtonen [1]

1 Department of Electrical Engineering and Automation, Aalto University, 02150 Espoo, Finland; andreas.lund@aalto.fi (A.L.); mahdi.pourakbari@aalto.fi (M.P.-K.); matti.lehtonen@aalto.fi (M.L.)
2 Granlund Consulting Ltd., 00700 Helsinki, Finland
* Correspondence: arslan.bashir@aalto.fi

**Abstract:** To achieve a successful integration of fluctuating renewable power generation, the power-to-heat (P2H) conversion is seen as an efficient solution that remedies the issue of curtailments as well as reduces carbon emissions prevailing in the district heating (DH) sector. Concurrently, the need for storage is also increasing to maintain a continuous power supply. Hence, this paper presents a MILP-based model to optimize the size of thermal storage required to satisfy the annual DH demand of a community solely by P2H conversion employing renewable energy. The DH is supplied by the optimal operation of a novel 2-km deep well heat pump system (DWHP) equipped with thermal storage. To avoid computational intractability, representative time steps with varying time duration are chosen by employing hierarchical agglomerative clustering that aggregates adjacent hours chronologically. The value of demand response and the effect of interannual weather variability are also analyzed. Numerical results from a Finnish case study show that P2H conversion utilizing small thermal storage in tandem with the DWHP is able to cover the annual DH demand, thus leading to a carbon-neutral DH system and, at the same time, mitigating the curtailment of excessive wind generation. Compared with the annual DH demand, an average thermal storage size of 29.17 MWh (2.58%) and 13.99 MWh (1.24%) are required in the business-as-usual and the demand response cases, respectively.

**Keywords:** power-to-heat; sector coupling; thermal storage; district heat; deep well heat pump; hierarchical agglomerative clustering; chronology; demand response; two-capacity building model



**Citation:** Bashir, A.A.; Lund, A.; Pourakbari-Kasmaei, M.; Lehtonen, M. Optimizing Power and Heat Sector Coupling for the Implementation of Carbon-Free Communities. *Energies* **2021**, *14*, 1911. https://doi.org/10.3390/en14071911

Academic Editor: Dimitrios Katsaprakakis

Received: 18 February 2021
Accepted: 26 March 2021
Published: 30 March 2021



## 1. Introduction

The current energy policies of the European Union (EU) and the Intergovernmental Panel on Climate Change (IPCC) of the United Nations (UN) urge combating global warming. As a result, the share of renewable energy generation capacity is substantially increasing in energy systems. A key strategy is to replace centralized fossil fuel-based energy generation with widely distributed clean renewable sources. This points to a massive energy transition ahead, including long-term planning of generation expansion and storage capacity. This clean energy transition not only includes the electrical power sector but also necessitates more actions in the district heating (DH) sector, as most of the end-use energy in Europe is in the form of heating [1]. For instance, heating and domestic hot water, together, constitute over 80% of the households' end-use energy in Finland [2]. Moreover, DH is mostly based on fossil fuels in Northern Europe such that it stands for 51% of all emissions in Helsinki, the capital of Finland [3].

While renewable energy sources (RES), such as wind and solar power, play an important role in the deep decarbonization of energy systems, they come with inherent variable and intermittent power generation levels that render them un-dispatchable. Thus, in order to achieve a successful integration of such RES technologies, smart grid solutions become apparent. In this field, sector coupling of electricity and heat utilizing excess renewable power generation to perform power-to-heat (P2H) conversion is an extensively proposed solution and has received increasing interest [4]. It implies that the excess renewable generation, which would otherwise be curtailed, can be used to produce DH. The benefit of this idea is twofold: first, it reduces the generation curtailments, and second, it reduces the carbon emissions in the DH sector. In simple words, the DH system has a great tendency to act as a sink for fluctuating RES generation.

Growing efforts have been dedicated to analyzing the flexibility and economic value associated with P2H coupling. For example, it was concluded in [5] that P2H conversion utilizing renewable generation was the most cost-efficient approach for minimizing carbon emissions for the case of the Nordic–Baltic region. The results also highlighted the need for thermal energy storage in conjunction with P2H application. Likewise, the authors in [6] demonstrated that utilizing only heat pumps would significantly reduce the emissions of DH in Helsinki under high wind power penetration. The work in [7] considered a 50% share of wind generation for Helsinki and established that the P2H option complemented with thermal storage would eliminate all the wind power curtailments. In [8], by considering national and local cases in Finland, it was argued that electrification of the heating sector could prove to be effective in mitigating $CO_2$ but with a warning that wind generation would not easily replace the electricity cogeneration in DH systems, due to adequacy issues in peak hours. The foregoing discussed studies particularly focused on the Nordic countries, as the DH load is quite significant due to the extreme and long winter season in this region. Moreover, thermal storage was emphasized as an implicit component in P2H application to eradicate generation curtailments.

While there is a consensus on the benefits of P2H coupling in mitigating curtailments, demand response (DR) is also characterized as a cost-effective tool to address RES integration. DR aims to provide upward or downward regulating power to balance the intermittent power levels of RES. Among DR loads, thermostatically controlled loads (TCLs), such as heating, ventilation and air-conditioning and domestic hot water usage, etc., occupy a leading slot. In the case of TCLs, the user-comfort is a function of temperature dead band only, while a smart thermostat is a fundamental requirement to control the thermal comfort level. The promising benefits offered by TCLs in the context of network management, investment planning and balance reserves have gained much attention. For instance, the system-wide potential of space heating loads for capacity planning of hydrogen energy storage in a highly renewable energy case of the Finnish power system was considered in [9]. In [10], the flexibility potential of space heating together with domestic hot water loads was analyzed, aiming at mitigating renewable generation curtailments for a residential microgrid. Similarly, the on-site utilization of solar energy generation was optimized by the support of TCLs in [11]. The above-discussed studies centered on space heating loads also considered building thermal dynamics for efficient DR coordination.

In our previous work [12], the P2H option was considered, along with the DR of space heating in a short-term operation-planning model enabling aggregator and households to communicate with each other to achieve mutual benefits. However, in order to account for events concerning high wind days or even very low wind days in P2H application for DH demand, thermal storage is certainly required, as endorsed in the literature. Therefore, our current work aims at optimal capacity planning of thermal storage in a DH system relying entirely on P2H conversion to achieve a zero-emission system.

Indeed, the capacity planning of energy systems is a long-term planning problem spanning several years, and it needs to consider short-term operating conditions, as well [13]. It implies using hourly or sub-hourly time representation of the planning horizon, which renders the model computationally intractable [14]. To deal with such issues, temporal complexity reduction techniques, also known as time period clustering, are often applied to power system planning studies. The goal of such time period clustering techniques is to enhance computational efficiency, while preserving as much information as possible. Consequently, a smaller number of representative periods is chosen to represent the whole horizon. One of the broadly accepted approaches requires representing the time-dependent parameters using system states. For instance, demand is first represented using duration curves. The duration curves are segregated into blocks, which are approximated with a finite number of segments [15]. Although this representation makes the model computationally efficient, a major drawback is that it totally neglects the chronology and, therefore, does not enable the inclusion of time-linking constraints, such as storage state of charge dynamics, tracking indoor thermal comfort of end-users and generation ramping constraints in the capacity planning model.

A more widely accepted time period clustering routine involves choosing representative days or weeks from a planning horizon to represent the annual variability of input data [16,17]. Contrary to the duration curves, this clustering approach is capable of maintaining some chronology, at least within each representative day or week, which also allows consideration of intraday time-linking constraints. In the context of storage investment, the authors in [18] proposed a representative time period model with a transition matrix to introduce some continuity between representative periods. The model was shown to outperform both system state and conventional representative time period clustering. However, such methodology fails to accurately consider interday storage dynamics in a medium- or long-term planning horizon [14,19], which is crucial, particularly in highly renewable generation power systems. Moreover, electricity demand has a diurnal pattern that can be easily captured by representative periods. On the contrary, wind power generation is highly volatile. Such fast dynamics cannot be captured by representative days, and hence, these clustering approaches undermine the short-term operating conditions in long-term planning problems.

In this paper, we utilize the hierarchical agglomerative clustering procedure for retaining the chronological information of the planning horizon in order to accurately capture the interday storage dynamics and track end-user thermal comfort levels concerning space heating loads. This technique hierarchically clusters just adjacent hours chronologically, according to Euclidean norm. There exist only a few studies applying chronological time period clustering in power system expansion studies involving storage [20,21]. The authors in [20] carried out a comprehensive comparison among four planning models, namely the full hourly model (benchmark), a model based on 28 representative days, one based on four representative weeks, and one based on chronological clustering using 672 timeslots for storage expansion studies involving renewable generation. It was proved that the chronological clustering-based model outperforms the remaining two time period representations in terms of error of the objective value. More recently, in a similar context, the authors in [21] demonstrated a comparison between chronological clustering and coupling clustering for days. The chronological clustering was proved to work more efficiently in approximating the abrupt variations associated with wind. Thus, aggregating the consecutive hours chronologically, denoted here as chronological clustering, overcomes the drawbacks of classical clustering approaches and brings economic value in combining renewable generation with storage devices.

The contributions of this paper are summarized as follows:

I. A thermal storage planning model is proposed to cover the annual DH demand of a community solely by P2H conversion using wind power, thus leading to a fully carbon-free DH system.

II. Chronological clustering is applied to accurately capture the midterm dynamics of wind power and track the energy level of storage and thermal comfort level of each end-user, while keeping the proposed planning model computationally tractable.

III. The P2H conversion is performed using a novel 2-km deep well heat pump (DWHP) system. The thermal analysis of DWHP is conducted separately in Finnish conditions over a 25-year study period. Moreover, the DWHP is used for both heat extraction and heat injection.

IV. Space heating load, thermal comfort and the associated DR are modeled using a two-capacity building thermal model.

V. The effect of interannual weather variability on storage size is studied by simulating the optimization model with different weather inputs.

The paper is structured as follows: In Section 2, a brief description of system models, including the hierarchical agglomerative clustering, is presented. Section 3 presents the optimization formulation for the thermal energy storage planning in DH systems employing P2H conversion through wind power. Section 4 discusses the simulation results for a Finnish case based on the EU visions for the years 2030 and 2050 [22]. Finally, Section 5 concludes the paper.

## 2. System Description

### 2.1. Two-Capacity Building Model

The two-capacity building model is accurate enough to estimate the space heating load of a detached house or an apartment building [10,12,23]. The model evaluates the heating demand by capturing the indoor air dynamics relative to the outdoor air temperature variation. As the name asserts, it makes use of two thermal capacitances. One capacitance represents the building structure, $C^m$, while the other is designated to the indoor air $C^a$. There are two unknown temperatures, the indoor temperature $T^a$ and the building fabric temperature $T^m$, evaluating the demand for heating or cooling. In addition, other temperature parameters of the model comprise ventilation supply temperature $T^x$, outdoor temperature $T^e$ and ground temperature $T^g$. The temperature nodes are related to each other through heat conductance or heat capacity flow in case there is an airflow between nodes. It is assumed that there is no warming of the infiltrating air in the building structure $C^m$ and the infiltrating air temperature is the same as that of the external air. The windows are assumed to have a negligible thermal mass, compared to the building envelope. The heating or cooling power of the air-handling unit is convective in nature, and it is set to operate at a fixed temperature, $T^x$. $C^m$ is dominant relative to $C^a$, but $C^a$ has a vital role in analyzing indoor air dynamics. The model is portrayed in Figure 1.

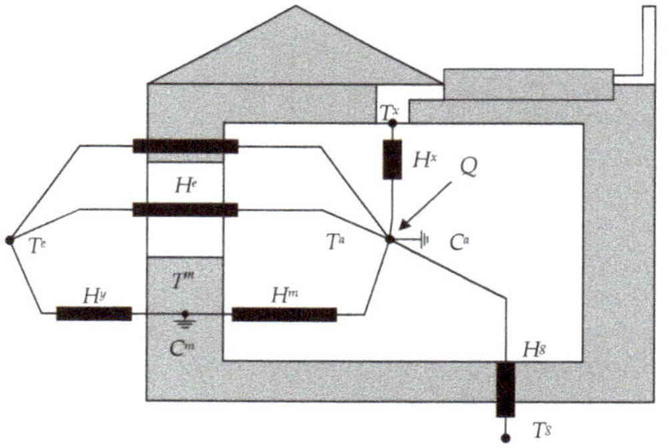


**Figure 1.** Two-capacity building model prototype [12].

The unknown parameters of the model are fine-tuned using the dynamic building energy simulation tool IDA–ICE, which is used to study the indoor climate and energy consumption of buildings. The test house is a single-family, two-floor house, with a total built floor area of 180 m$^2$. The structure of the house is medium-weight passive, according to the Finnish guidelines. More details of the prototype house, including the floor plan are given in [24]. During calibration, the indoor temperature set-point is 21 °C [25]. The heating power is cut off for six hours that results in an exponential drop in the indoor temperature. The parameters are determined by minimizing the variance between the response attained from the IDA–ICE model and the derived two-capacity model. The calibration procedure is performed at three different outdoor temperatures, i.e., −10 °C, 0 °C and 10 °C, as illustrated in Figure 2.

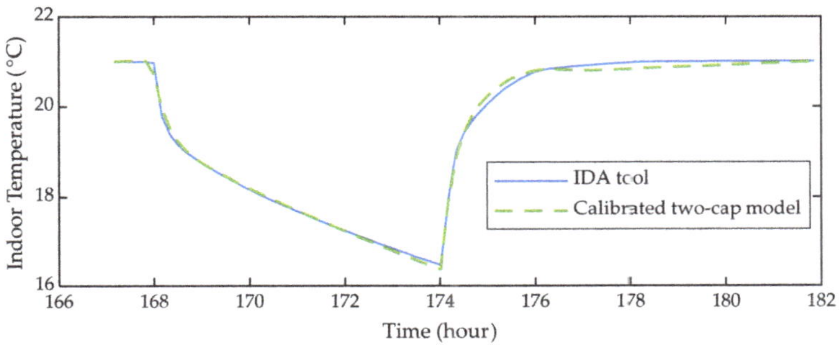


**Figure 2.** Calibration of the two-capacity building model [12].

### *2.2. Deep Well (Borehole) Heat Pump System*

The P2H conversion employing clean renewable energy can provide aid for the evolution of the DH sector, as well as offer new options for balancing the large-scale intermittent electricity generation. The P2H conversion is usually performed using community-scale electric boilers and small-scale heat pumps. Concurrently, cutting emissions also requires exploiting new types of heat sources and storages to match the scale of the DH consumption. Employing large-scale heat pumps utilizing geothermal energy can be a promising solution in the future. The geothermal energy extracted 100–300 m below the ground with stable temperatures between 5–7 °C is classified as primary geothermal energy, whereas the thermal energy extracted from the upper layers of the soil (a few meters deep), heated by solar radiation and ambient temperature, is the secondary geothermal energy. Here, the temperature varies from freezing conditions to 10–15 °C. In either case, a heat pump is necessary to increase the temperature of the heat source to the desired level, i.e., suitable for a DH system.

In most heat pump applications, geothermal energy utilization has been limited to primary geothermal energy. In Northern Europe and Finland, the geological conditions involve old and stable bedrock, where the upper ground layers have cooled down, and the thermal gradient in the rock is approximately 10–20 °C per km [26]. Hence, harnessing the geothermal energy for DH applications in Nordic conditions requires 1–3 km deep boreholes coupled with the heat pump. Such deep heat boreholes (or deep heat wells) have a temperature ranging between 20–40 °C. In Finland, most of the ground-coupled heat pumps exploit geothermal reserves less than 300 m deep, whereas deeper heat resources have not been exploited. There is one deep heat well, 7 km deep, under the piloting phase at the st1-Fortum site in Espoo, Finland, but the technology has not progressed for commercial use yet. Further, due to a difference in climate, geography and building code, the results obtained from deep heat wells installed elsewhere cannot be applied locally.

To date, there exist only a few studies addressing the thermal behavior of deep well heat exchangers. In this area, the authors in [27] carried out a field test of three 2-km deep heat wells situated in Xi'an, China. The results of the investigation were later matched with the results obtained by simulating the numerical model. Similarly, [28] developed a numerical model to study the thermal performance of a coaxial-shaped deep borehole heat exchanger and replicated the measurements obtained from distributed thermal response test from an installation in Norway. The simulation was done for boreholes in the depth range of 200–1000 m. Recently, a detailed simulation study on the thermal behavior of a 2-km deep heat well in Finnish conditions was accomplished in [29]. The deep heat well performance was analyzed for a period of 25 years, and it was shown to produce about 110 kW heat in a steady state. Moreover, the simulation model developed in [29] was validated using the results presented in [28].

In this work, we utilize the simulation results of the 2-km deep heat well analyzed in the Finnish geological conditions [29]. The reason for using the deep heat well is that its annual heat output is about 30 times more than the conventional 300 m deep well [29].

The deep well heat pump system (DWHP) is simply a combination of a deep heat well and a heat pump. The deep heat well consists of a coaxial tube having an outer and inner pipe incorporating opposite flows, as illustrated in Figure 3. The two pipes are insulated to avoid thermal short-circuiting, which contributes to thermal losses. The DWHP can be used for heat extraction and/or heat injection (thermal storage). In heat extraction mode, a fluid at low temperature enters the well along the outer pipe and exits through the inner pipe. During this flow, the fluid temperature increases, and this heat energy is transferred to the coupled heat pump, which further raises the temperature to the desired level. Contrarily, in heat injection mode, hot fluid enters the well through the middle pipe, transfers energy to the surrounding bedrock and flows back upwards as a cooled fluid through the outer pipe. Hence, the bedrock can be regenerated or charged, and the deep well operates as thermal storage. In both cases, pumping power for fluid flow and compression power for raising the fluid temperature are required, which can be utilized from wind power generation or any other renewable energy.

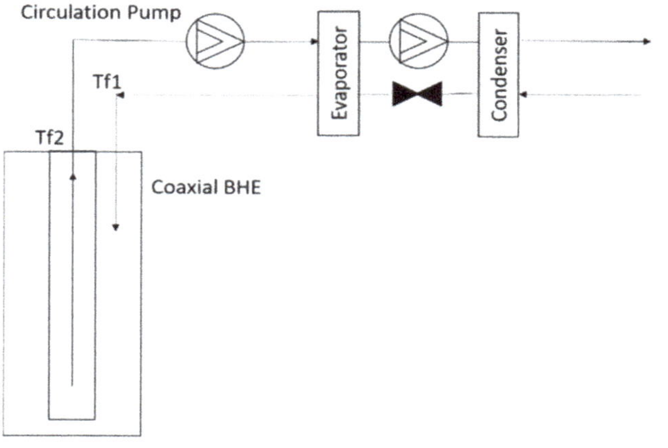


**Figure 3.** Coaxial borehole heat exchanger coupled with the heat pump.

The numerical model of the DWHP was developed in COMSOL Multiphysics (software version 5.4), and the thermal performance in Finnish geological conditions was simulated for a 25-year period with a time resolution of 7.5 h. A detailed description of the model, the physical parameters of the deep heat borehole and the COMSOL modules used can be found in [29,30]. In this work, the operation sequence of the DWHP is six months of heat generation, followed by six months of heat injection. This operation sequence is simulated for flow rates ranging from 0.5 kg/s to 15 kg/s to approximate the relationship between electrical power input and heat power output of the DWHP for both operation modes. Figure 4 illustrates the fluid temperature dynamics in borehole pipes for both cycles at a flow rate of 6 kg/s. Figure 4a portrays the heat extraction cycle where the fluid enters the outer pipe at 6 °C and leaves the borehole through the central pipe at a higher temperature, on average about 14 °C at the end of each extraction cycle. Contrarily, in Figure 4b, during the heat injection phase, hot fluid is injected at 70 °C through the central pipe, and the fluid returns through the outer pipe in the temperature range of 41–46 °C. Such a decline in temperature signifies that the surrounding bedrock absorbs heat. It can also be seen for both phases in Figure 4 that the thermal dynamics of the borehole are faster at the beginning, with a declining trend over the six months cycle. This saturation aspect is explained by the finite thermal conductivity of the bed rock. The temperature variation, depicted in Figure 4, can be converted into corresponding thermal powers.

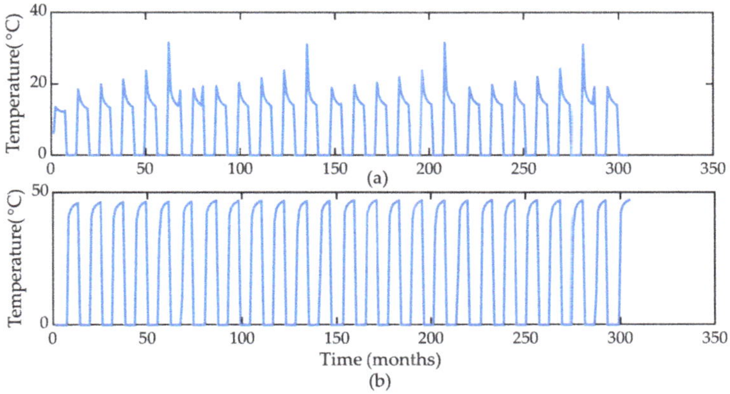


**Figure 4.** Thermal dynamics of deep well heat pump system (DWHP). (**a**) Fluid temperature in the central pipe during heat extraction. (**b**) Fluid temperature in the outer pipe during heat injection [12].

The performance features of the DWHP for heat extraction and injection cases are portrayed in Figure 5. Each characteristic curve is well-approximated using a quadratic regression curve. Note that the minimum input power refers to a fluid flow rate of 0.5 kg/s. The output heat power against each input in Figure 5 corresponds to the steady-state simulated value obtained in the last six-month cycle of the 25-year study period. Figure 5a shows that the extracted heat increases linearly until an input power of 77 kW. Increasing input beyond 77 kW reduces the system coefficient of performance (CoP) in the heat extraction mode. This is explained by the increased pressure drop when the flow rate is increased above 6 kg/s, resulting in larger hydraulic losses, as the losses are proportional to the square of the fluid velocity. In the heat injection phase, the heat absorption capacity of the borehole limits the injection power so that the amount of absorbed heat is always less than the available electricity input, as depicted in Figure 5b. The main limiting factor for the heat injection rate is the heat transfer from the fluid to the surrounding rock. The heat transfer inside the rock is dominated by the low heat conductivity, i.e., increasing the flow rate, and, thus, the convective heat transfer at the fluid–rock interface would not substantially improve the overall heat transfer rate.

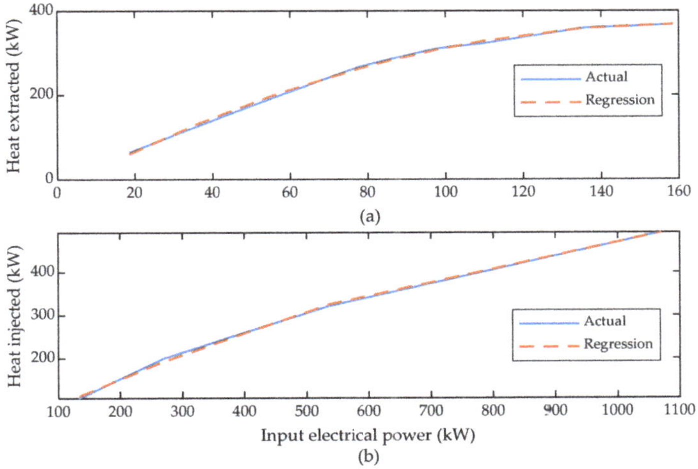


**Figure 5.** DWHP system operational characteristics. (**a**) Heat extraction. (**b**) Heat injection.

### 2.3. Time Series Aggregation

In the context of power system planning, time period clustering or aggregation is broadly used to shrink the planning horizon by choosing a reduced number of representative time periods (hours, days, weeks, etc.). Hierarchical clustering is used in the current subject matter, as the resultant representative time series is independent of the initialization of the clustering algorithm. In other words, the clusters are reproducible. Further, it allows the inclusion of additional principles regarding how individual clusters are iteratively merged to form new clusters. Additionally, it aggregates only consecutive hours chronologically. Unlike classical clustering approaches, hierarchical clustering uses a dissimilarity measure between each successive pair of clusters to determine which clusters are merged in hierarchical agglomerative clustering. The hierarchical agglomerative clustering used here is centered on Ward's linkage criterion. Ward's method involves recursively merging consecutive pairs of clusters that minimally increase within cluster variance [31]. The clustering procedure used in this work is explained below.

At first, the time-dependent parameters are normalized, so that the clustering would be unaffected by the different orders of magnitude of parameters. Then, the time series of all parameters are concatenated in a single time series $\{X_k\}_{k=1,\ldots,8760}$. For example, in our case, there are two time-dependent parameters, namely outdoor temperature $T_k^e$ driving the space heating load and wind speed $w_t$, to determine the wind power generation, $X_k = (T_k^e, w_k)$. Let $R$ and $R'$ be the initial and the desired number of time steps (clusters) of the normalized time series $X_k$, respectively. The steps to arrive from $R$ to $R'$ clusters are listed below:

I. The algorithm initializes as every single entry of the time series as its own cluster. Fix the initial number of clusters $r$ to the total number of time steps, i.e., 8760, in our case.

II. Compute the centroid of each cluster $K$ according to Equation (1)

$$\overline{X}_K = \frac{1}{|K|}\sum_{k\in K} X_k \quad (1)$$

III. Evaluate the dissimilarity index between each pair of contiguous clusters $K$ and $L$ according to Ward's criterion, as in Equation (2). $|K|$ denotes the number of hours in cluster $K$.

$$D(K,L) = \frac{2|K||L|}{|K|+|L|}\left\|\overline{X}_K - \overline{X}_L\right\|^2 \quad (2)$$

IV. Based on the dissimilarity matrix, combine the two closest adjacent clusters to form a new cluster ($K'$, $L'$); ($K'$, $L'$) argmin $D(K, L)$, where $L$ belongs to the set of clusters adjacent to the cluster $K$.

V. Update $r$ with $r - 1$.

VI. If $r = R'$, proceed to step VII; otherwise go to step II.

VII. Replace the clusters formed in $X_k$ with their cluster centroids. The duration of each time step (cluster) in the representative time series is equal to the number of hours forming that cluster.

VIII. The times series is rescaled to obtain the absolute values.

Note that the chronological clustering reduces the number of time-dependent parameters, variables and equations from $R$, i.e., 8760, to $R'$. For visualization, Figure 6 illustrates the aggregation of wind speed (p.u.) time series performed using chronological clustering. For a clear depiction, 200 consecutive hours are chosen out of the 8760 h clustered using only 1752 time-steps, i.e., one-fifth of the actual time series. Each candidate cluster is represented by its centroid. It can be seen in Figure 6 that time steps have a variable time duration and the methodology well-approximates the highly fluctuating wind speed profile. Such time series cannot be accurately aggregated using the methodology employing representative weeks, days or hours. In the case of multiple time-dependent parameters hierarchical clustering is applied to all parameters simultaneously in blocks. In this work the weather parameters, namely wind speed and outdoor temperature, are clustered using the method described above.

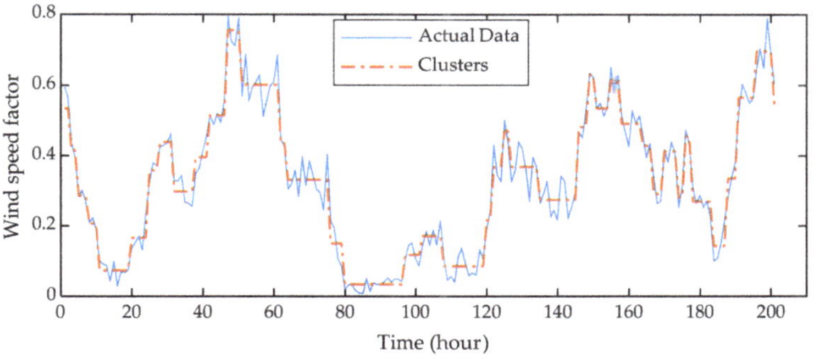


**Figure 6.** An illustration of chronological clustering using wind speed time series.

## 3. Optimization Model

This section presents the mathematical model for optimal sizing of the thermal storage required to satisfy the DH demand of a community by P2H schemes. The storage sizing model is formulated as a mixed-integer linear programming (MILP) model. To keep the model simple, it is formulated as a deterministic optimization problem and planned for a single year. However, the model was tested with different input time series to study the interannual weather variability. The model is presented below.

$$Minimize\ SOC_{\max}^{TS} \tag{3}$$

$$P_t^{w} - P_t^{wc} = P_t^{TS,ch} + P_t^{dwhp,e} + P_t^{dwhp,i} \tag{4}$$

$$Q_t^{dwhp,e} + Q_t^{TS,dch} = \sum_n Q_{t,n}^{h} \tag{5}$$

$$SOC_t^{TS} = SOC_{t-1}^{TS} + \eta^{ch} P_t^{TS,ch} \omega_t - Q_t^{TS,dch} \omega_t / \eta^{dch} - \varepsilon_t^{loss} \tag{6}$$

$$\varepsilon_t^{loss} = \mu\, SOC_{t-1}^{TS}\, \omega_{t-1} \tag{7}$$

$$0 \le P_t^{TS,ch} \le P_{\max}^{TS,ch} \tag{8}$$

$$0 \le Q_t^{TS,dch} \le Q_{\max}^{TS,dch} \tag{9}$$

$$SOC_{\min}^{TS} \le SOC_t^{TS} \le SOC_{\max}^{TS} \tag{10}$$

$$SOC_{t=end}^{TS} \ge SOC_{t=t_0}^{TS} \tag{11}$$

$$\sum_t Q_t^{BH,i} \omega_t = \sum_t Q_t^{BH,e} \omega_t \tag{12}$$

$$Q_t^{dwhp,e} = A_1 \left(P_t^{dwhp,e}\right)^2 + A_2 P_t^{dwhp,e} + A_3 u_t \tag{13}$$

$$Q_t^{BH,e} = B_1 \left(P_t^{dwhp,e}\right)^2 + B_2 P_t^{dwhp,e} + B_3 u_t \tag{14}$$

$$P_{\min}^{dwhp,e}\, u_t \le P_t^{dwhp,e} \le P_{\max}^{dwhp,e}\, u_t \tag{15}$$

$$P_t^{dwhp,e} = \sum_i D_i y_{t,i} \tag{16}$$

$$\left(P_t^{dwhp,e}\right)^2 = \sum_i (D_i)^2 y_{t,i} \tag{17}$$

$$\sum_i y_{t,i} = 1 \tag{18}$$

$$Q_t^{BH,i} = C_1\left(P_t^{dwhp,i}\right)^2 + C_2 P_t^{dwhp,i} + C_3 v_t \tag{19}$$

$$P_{\min}^{dwhp,i} v_t \le P_t^{dwhp,i} \le P_{\max}^{dwhp,,i} v_t \tag{20}$$

$$P_t^{dwhp,i} = \sum_j E_j z_{t,j} \tag{21}$$

$$\left(P_t^{dwhp,i}\right)^2 = \sum_j \left(E_j\right)^2 z_{t,j} \tag{22}$$

$$\sum_j z_{t,j} = 1 \tag{23}$$

$$u_t + v_t \le 1 \tag{24}$$

$$T_{t,n}^a = \frac{T_{t-1,n}^a + \frac{\omega_t}{C_n^a}\left(H_n^m T_{t,n-1}^m + H_n^e T_t^e + H_n^g T_n^g + H_n^x T_n^x + \phi_{t,n}^{sol} + \phi_{t,n}^{\text{int}} + Q_{t,n}^h\right)}{1 + \frac{\omega_t}{C_n^a}\left(H_n^m + H_n^e + H_n^g + H_n^x\right)} \tag{25}$$

$$T_{t,n}^m = \frac{T_{t-1,n}^m + \frac{\omega_t}{C_n^m}\left(H_n^m T_{t-1,n}^a + H_n^y T_t^e\right)}{1 + \frac{\omega_t}{C_n^m}\left(H_n^m + H_n^y\right)} \tag{26}$$

$$T_n^{set} - \frac{\Delta}{2} \le T_{t,n}^a \le \left(T_n^{set} + \frac{\Delta}{2}\right) + M\, x_{t,n}^h \tag{27}$$

$$0 \le Q_{t,n}^h \le Q_n^{\max}\left(1 - x_{t,n}^h\right) \tag{28}$$

$$\sum_t Q_{t,n}^h \omega_t = \beta_n^h \tag{29}$$

$$\sum_t P_t^{wc} \omega_t \le \alpha \sum_t P_t^w \omega_t \tag{30}$$

In the above problem, the objective function (Equation (3)) is the minimization of thermal storage capacity. The electrical power balance and the DH power balance are managed in Equations (4) and (5), respectively. Note that all the DH loads must be satisfied using wind power in order to ensure zero emissions in the DH sector. The thermal storage charging and discharging dynamics are captured in Equation (6). The thermal storage is charged using electricity from wind power generation, where it is stored as heat energy. Due to the nonideal behavior of the storage, some stored heat energy would be lost. Such losses are modeled as a linear function of stored energy weighted with a loss coefficient in Equation (7). The thermal storage charging and discharging powers are capped in Equations (8) and (9), respectively. Constraint (10) states that the stored thermal energy can hover within specified limits only, while Equation (11) guarantees that the final energy level of the thermal storage is greater than or equal to the initial level. Constraint (12) maintains the energy balance of the deep borehole by ensuring that the total heat energy extracted from the borehole is equal to the heat energy injected over a yearly operating cycle. Although the borehole reaches a steady-state operation in 10–15 years of continuous operation, maintaining such energy balance is quite necessary for long-term operation. Constraint (13), using the regression curve depicted in Figure 5, relates the heat output with the electricity input of the DWHP system, i.e., the borehole coupled with heat pump, whereas Equation (14) relates the DWHP input with the extracted heat of the deep borehole, alone. The input limits of the DWHP during heat extraction are defined in Equation (15). The binary variable in Equations (13)–(15) ensures that the DWHP operates according to specified limits, as depicted in Figure 5.

In order to linearize the quadratic term in Equations (13) and (14), special ordered sets of type 2 (SOS2) variables are employed in Equations (16) and (17). SOS2 is a set of variables in which at most two variables take nonzero values, and these variables are adjacent to each other [12,32]. The base of the quadratic term, i.e., a continuous variable, is first represented as a sum of a finite number of breakpoints weighted with SOS2 variables, as in Equation (16). The square of the continuous variables is then evaluated using Equation (17). Another necessary condition is that the sum of SOS2 variables must be equal to unity, which is implemented in Equation (18). The Constraints (19) and (20) describe the operation of DWHP in heat injection mode following the same linearization procedure as in Equations (21)–(23). To ensure that both heat extraction and heat injection do not occur simultaneously, Equation (24) is applied.

Constraints (25) and (26) represent the discrete version of the two-capacity building model that is used to calculate the space heating demand and capture the indoor ambient temperature variation in a detached house. When DR is enabled, the indoor temperature can mutate between a predefined dead band, as modeled in Equation (27). A binary variable is used to relax the upper comfort limit of the indoor temperature if the day gets warm and heating is not needed [33]. It is further supported by Constraint (28), establishing that heating is either zero or turned on to keep the indoor temperature within the thermal comfort band. According to Equation (29), the net flexibility of space heating load for each household is set to zero over the horizon, so that the annual demand remains the same with or without activating DR. Finally, the allowed wind curtailment level can be tuned using Equation (30) to ensure that the thermal storage operation coordinates with the available wind generation.

## 4. Case Study

### 4.1. Simulation Parameters

We consider the weather profile of Helsinki, Finland. The wind speed time series $P_t^{w}$ at the height of 50 m above sea level was prepared using the framework in [34]. A statistical approach was employed by [34] without any site-specific measured data to simulate 100 yearly wind speed profiles at varying heights for different locations in Finland. The hourly outdoor temperature profile was obtained from the Finnish Meteorological Institute [35]. Having known the wind speed, the corresponding wind power generation could be computed using cut-in speed and rated speed of wind turbine, which can be obtained from the wind turbine datasheet. We considered a wind farm with a rated output power of 700 kW, which was increased in steps. For the DWHP operation, SOS2 variables with three breakpoints were used in both the heat extraction and the injection case.

The time-dependent weather parameters spanning 8760 h were clustered using the methodology explained in Section 2.3. Figure 7 illustrates the wind speed and the outdoor temperature profiles using 1252 representative time steps obtained by chronologically clustering the 8760 h. Although, in Figure 7, the time resolution is not hourly, based on the duration of each time step, abrupt variations and the seasonal trend in the input profiles are still appropriately captured by chronological clustering. The reason to cluster such driving factors directly instead of corresponding power levels is that the parameter, such as outdoor temperature, significantly affects the space heating decision variable, thus enables to utilize the clustered data in the DR mechanism separately.

A small-scale community comprising 100 single-family two-floor detached houses, each equipped with the DH facility, was considered. It was assumed that there was an aggregator that owned a wind farm and controlled the operation of the DWHP system and a separate thermal storage in tandem. Moreover, a smart environment was assumed, in which the aggregator was authorized to schedule the DH load of each house and knew the individual load parameters, including thermal comfort levels, by communicating with the home energy management system (HEMS) installed with each end-user. The aggregator operated the resources optimally to minimize the thermal storage size, while covering the DH demand through P2H conversion, only. The calibrated two-capacity model parameters of the prototype house are listed in Table 1 [11,12]. The rated power of the heating unit for each house was distributed around a mean value of 7 kW, while the mean house area was 180 m$^2$. The indoor heating set-point temperature of each household was 21 °C, which could deviate within ±0.5 °C when the DR was activated [25]. To reduce the computational burden further, a set of fewer representative houses with the DH load comprising only space heating load was considered in this study. The domestic hot water consumption was ignored, as well as that the associated DH network losses were negligible. It was assumed that houses were continuously occupied over the study horizon so that there was always a heating demand, and thermal comfort had to be maintained in each time step of the horizon. The heat gains from lighting, equipment, occupants and solar irradiation in a detached house were simulated using the IDA–ICE tool. Such heat gains for a typical house have been depicted in Figure 8. Based on these assumptions, Figure 9 illustrates the corresponding space heating demand and wind power generated using the clustered data shown in Figure 7. The annual DH demand and the wind power production corresponding to the given weather profiles total 1.13 GWh and 2.05 GWh, respectively.

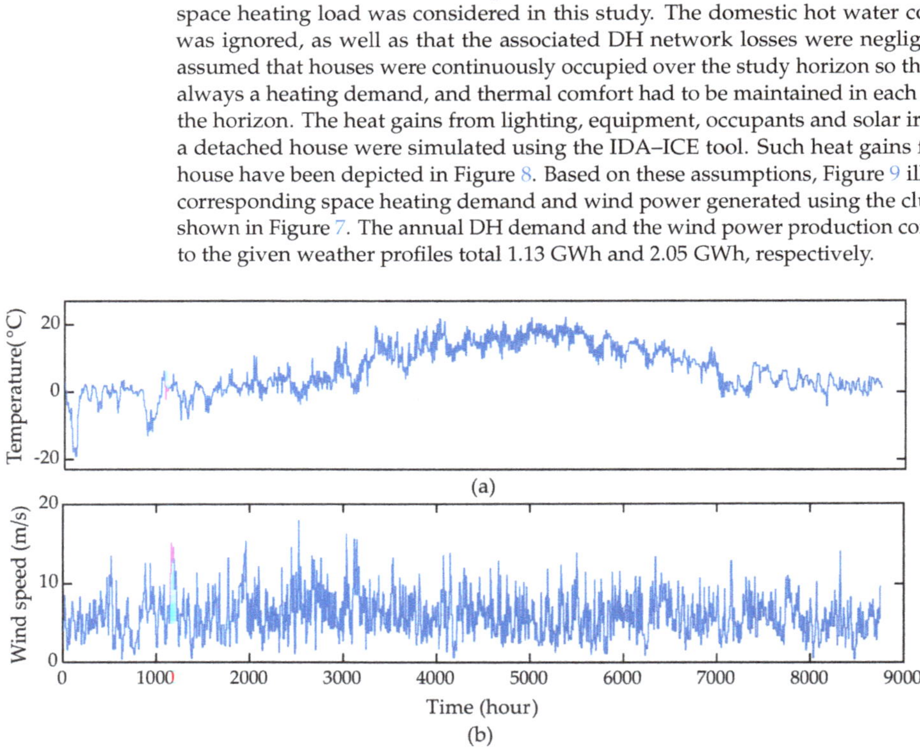


**Figure 7.** Input data time series using 1252 representative time steps clustered chronologically. (**a**) Outdoor temperature. (**b**) Wind speed at 50 m height.

**Table 1.** Calibrated two-capacity model parameters.

| Parameter | Unit | Value |
|---|---|---|
| $H^e, H^m, H^g, H^x, H^y$ | W/°C/m$^2$ | 0.29, 5.16, 0.05, 0.48, 0.33 |
| $C^m, C^a$ | Wh/°C/m$^2$ | 31.14, 3.616 |
| $T^x, T^g$ | °C | 18, 10 |

The simulation horizon was one year, i.e., from 1 January to 31 December. The input data included outdoor temperature, wind speed, two-capacity building parameters, households' thermal comfort preferences and their load parameters. The decision variables comprised thermal storage size, its charging and discharging power, capacity variation, curtailed wind power, the DWHP operation and the DR. The chronological clustering was performed in Matlab prior to simulating the MILP problem (Equations (3)–(30)) in Matlab–GAMS platform. The CPLEX solver was used to solve the proposed model on a Windows desktop computer with a 3.4 GHz Intel Xeon processor and 48 GB RAM. The simulation time related to clustering, for reducing 8760 h to 1252 representative time steps was about 55 min, whereas the simulation time of the MILP problem (Equations (3)–(30)) employing 1252 representative time steps was about 7 min.

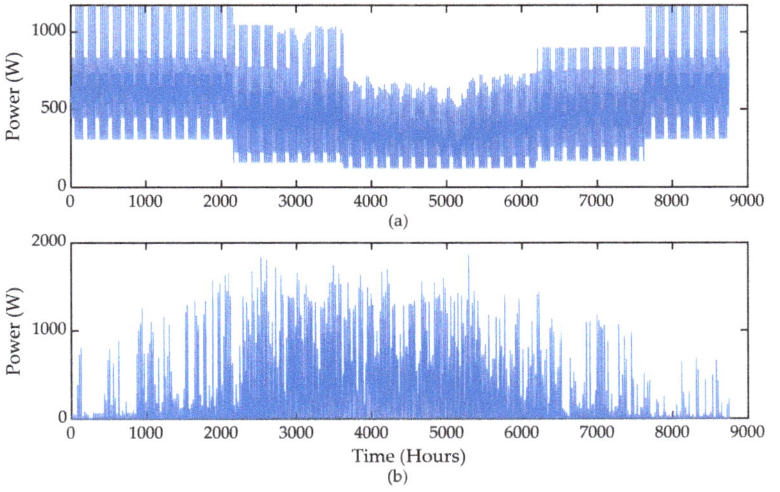


**Figure 8.** Heat gains in a detached house. (**a**) Internal heat gains from lighting, equipment and occupants. (**b**) Solar heat gains.

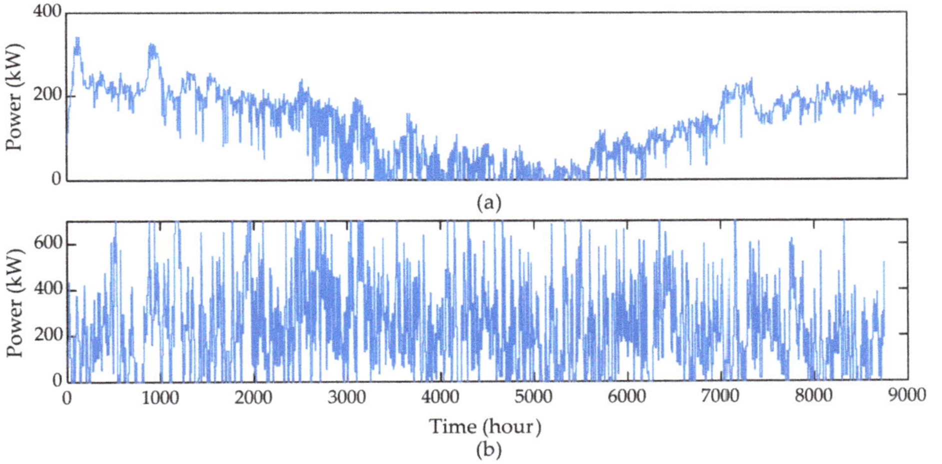


**Figure 9.** Input data time series using 1252 representative time steps. (**a**) Total district heat demand. (**b**) Wind power.

### 4.2. Simulation Results

We simulated two cases. The first was the business-as-usual (BAU) case, in which the indoor temperature of each household was strictly fixed to the set-point temperature of 21 °C during the heating period. It was accomplished by setting the thermal comfort band to zero in Equation (27). In the second case, DR was activated, and the indoor temperature could vary within ±0.5 °C from the set-point. For either case, wind curtailment level was set to 10% at the most, and the thermal storage parameters were set as $\eta^{ch} = \eta^{dch} = 0.9$, initial SoC was 50% and loss coefficient $\mu$ = 0.2% per hour.

Figure 10 demonstrates the thermal storage evolution for the BAU case. The SoC dynamics of the thermal storage show that its operation was more demanding in the winter season, when the DH demand was also high. On the contrary, storage was occasionally charged or discharged during the summer season, since the heating was turned on and off, depending on the outdoor temperature, as depicted in Figure 9a. The optimal storage size was simulated as 31.44 MWh, which just forms 2.78% of the annual DH demand. Such an outcome is significantly supported by coordinating the DWHP operation with the thermal storage. The heat extracted and injected from the deep heat borehole is depicted in Figure 11. For a comparison, heat extracted from the DWHP is also portrayed alongside. In Figure 11, the difference between the heat extracted from the borehole, alone, and that from the DWHP system represents the heat pump compression power required to further raise the temperature of the heat source. Moreover, heat injection mostly functions during low-demand periods, when thermal storage is also oscillating around its maximum capacity. Simulation results show that a total of 337.3 MWh of heat energy was extracted from or injected into the borehole, alone.

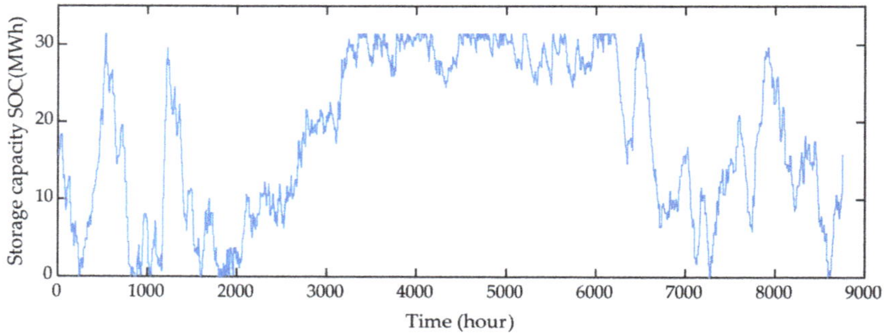


**Figure 10.** State of charge (SoC) of the thermal storage in the business-as-usual (BAU) case.

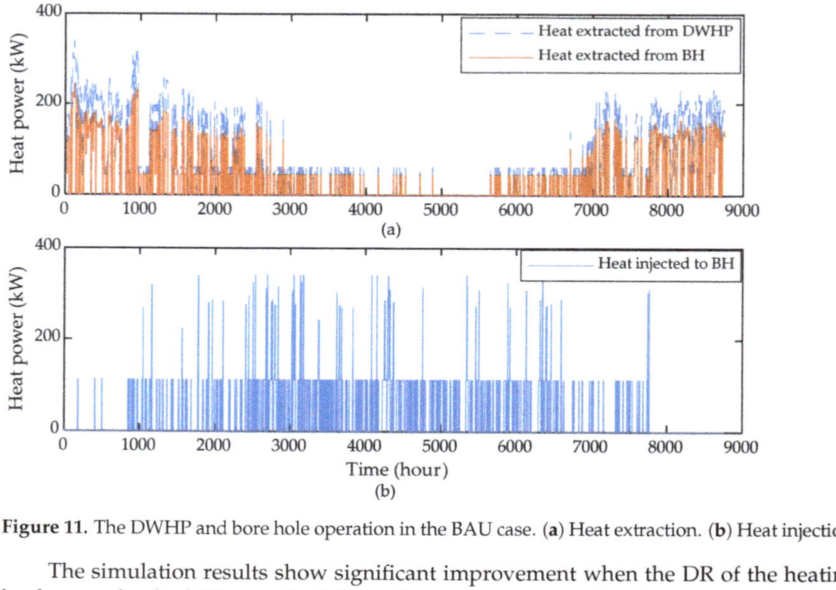


**Figure 11.** The DWHP and bore hole operation in the BAU case. (**a**) Heat extraction. (**b**) Heat injection.

The simulation results show significant improvement when the DR of the heating load was unleashed. The required thermal storage capacity reduced to 12.87 MWh, which is 2.5 times smaller, as compared to that in the BAU case. Moreover, this storage size merely constitutes 1.14% of the annual DH demand. The storage capacity variation over the horizon is illustrated in Figure 12. The DR case results in more heat energy extracted or injected into the bore hole, as depicted in Figure 13, and this energy aggregates to 482.45 MWh per annum. The reason behind this outcome is that the maximum allowed wind power curtailment level was maintained via Equation (30). The BAU case had more opportunities to utilize wind power, due to a higher thermal storage capacity and storage losses, which required more charging energy during the horizon. Contrarily, the thermal storage in the DR case resulted in a relatively smaller capacity and, hence, led to lower storage losses and the charging energy. To maintain the desired curtailment level, more wind power was utilized to operate the DWHP system, instead.

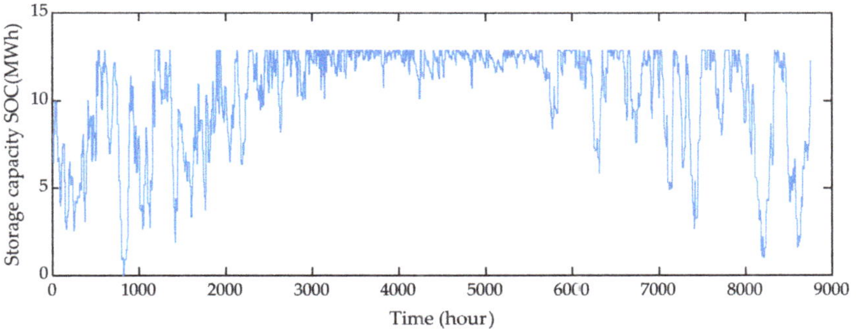


**Figure 12.** State of charge (SoC) of the thermal storage in the demand response (DR) case.

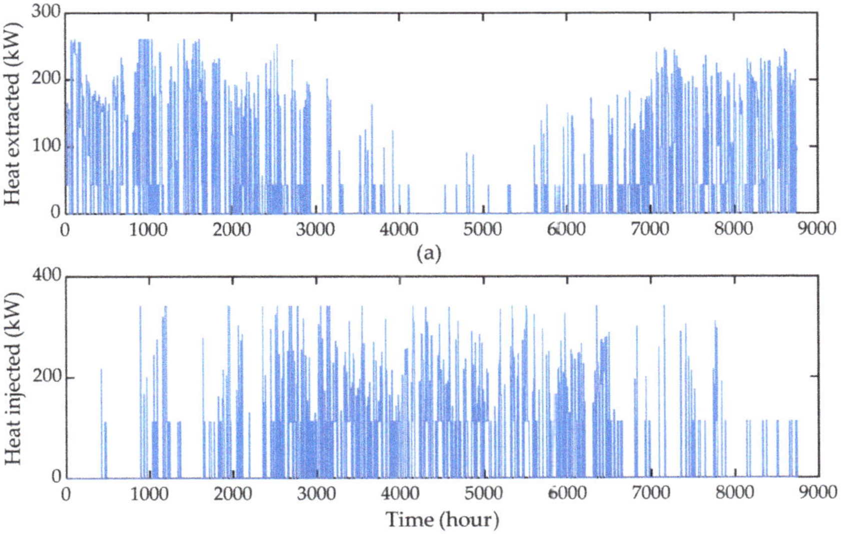


**Figure 13.** The deep heat bore hole operation in the DR case. (**a**) Heat extraction from borehole. (**b**) Heat injected to borehole.

Figure 14 shows the DH demand during DR, which is quite different from that in the BAU case, as depicted in Figure 9a. In Figure 14, the DH demand has a fluctuating nature, as the heating units are frequently operated to utilize the extreme limits of the predefined thermal comfort band to effectively balance the wind generation level. Nonetheless, the annual aggregated demand remained the same, i.e., 1.13 GWh. Note that, according to the simulation results, even in the absence of a heating period, the indoor ambient temperature remained between 21 °C and 22 °C for both cases.

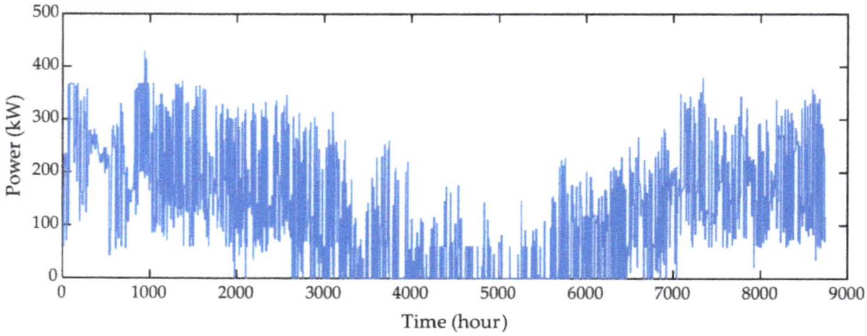


**Figure 14.** District heating demand in the DR case.

Figure 15 demonstrates the variation in thermal storage size and the wind power spillage with respect to the increasing wind power penetrations. Please note that higher wind power penetrations require higher wind power curtailment limits if the DH demand is the same. For this reason, the wind curtailment limit was increased to 40% to carry out this sensitivity analysis. As can be seen in Figure 15, the required storage capacity remarkably decreases from 2.78% of the aggregated DH demand to 1.7% when the wind generation is oversized to about 140% of the initial capacity in the BAU case. Contrarily, in the DR case, the storage size did not show any significant improvement with the increasing wind power penetrations and rather showed a saturating trend. Figure 15 also proves that the DR of the DH demand is more effective in reducing the storage size when compared with increasing wind power investments, which also results in more wind power spillage.

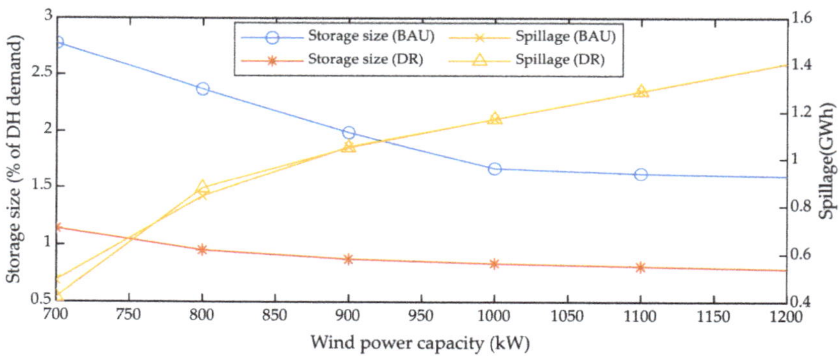


**Figure 15.** Effect of increasing wind power penetration on storage size and wind generation curtailments (Maximum wind curtailment = 40%).

The simulation results presented above were based on a single run of wind speed time series. To account for the variability and uncertain nature associated with the weather parameters, it was important to solve the optimization model for different input data and determine relevant statistics of the results. For this purpose, the model (Equations (3)–(30)) was solved for 19 distinct weather input time series obtained from [34]. Each weather time series data was first clustered chronologically into 1252 representative time-steps, and the model was simulated for both the BAU and the DR case. The mean, standard deviation and confidence bounds were determined, which are summarized in Table 2. The spread, including the interquartile range of the obtained results, are also demonstrated using a box-plot in Figure 16. The higher standard deviation relative to the mean value for the DR case, as listed in Table 2, is explained by the higher extreme values and the outlier, as can be seen in Figure 16. However, the mean value falls within the interquartile range, implying that considering a greater number of inputs would certainly reduce the standard deviation and shorten the interquartile range, as well.

**Table 2.** Storage size statistics.

| Case | Mean (MWh) | Standard Deviation (MWh) | Lower 95% Confidence Bound (MWh) | Upper 95% Confidence Bound (MWh) |
|---|---|---|---|---|
| BAU | 29.172 | 7.66 | 25.72 | 32.61 |
| DR | 13.99 | 5.37 | 11.58 | 16.40 |

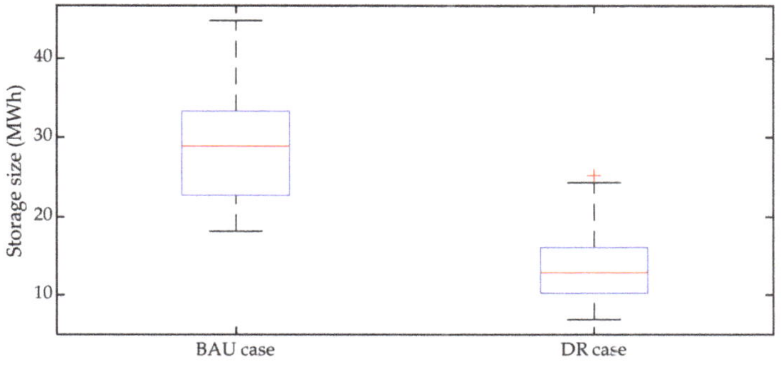


**Figure 16.** Box-plot representation of the simulation results for various wind speed profiles.

Lastly, the impact of increasing the representative periods on the energy storage size was investigated. Table 3 lists the outcome when the model (Equations (3)–(30)) was simulated with a higher number of representative time periods. The effect on the storage size was negligible when representative periods were increased from 1252 to 2920, but the simulation time increased rapidly. The simulation time nearly doubled at each stage, when more details of the time series were added, as is visible in Table 3. The model became intractable before the number of representative periods reached one-half of the number of hours in the actual time series.

**Table 3.** Effect of increasing the representative time periods.

| No. of Representative Time Periods | Time Series Representation (×8760) | Storage Size (MWh) | Simulation Time |
|---|---|---|---|
| 1252 | 1/7 | 12.87 | 6 min 46 s |
| 1460 | 1/6 | 12.53 | 14 min 38 s |
| 1752 | 1/5 | 12.81 | 32 min 39 s |
| 2190 | 1/4 | 13.27 | 1 h 49 s |
| 2920 | 1/3 | 13.32 | 2 h 16 min |

## 5. Conclusions

Renewable energy sources can play a decisive role in reducing carbon emissions that are proliferating in the district heating sector. However, the inherent intermittency of such energy sources offers a significant integration challenge to the current power system. This work was aimed at mitigating curtailments from the perspective of power systems and simultaneously concluded a zero-emission district heating system by utilizing direct P2H conversion from wind generation. The task utilizes a 2-km-DWHP system and thermal storage coupled with the existing DH system. Accordingly, a framework was proposed to minimize the size of the community-scale thermal storage. The second contribution involved employing chronological time period clustering to accurately solve the optimization problem for a period of one year. The model was applied to a case study in Finland. The main findings are summarized as follows:

- The simulation results proved that considering a small thermal storage in tandem with the DWHP prototype can cover the DH demand of a community. On average, the required optimal size of the thermal storage in the BAU case is just 29.172 MWh, i.e., 2.58% of the DH demand.
- DR can play a complementary role to economically satisfy the DH demand without oversizing the wind generation, which would otherwise result in an increased power spillage. Activating DR reduces the storage size, on average, to 13.99 MWh, i.e., 1.24% of the DH demand.
- The precision of the results was evaluated by increasing the number of representative time periods. The storage size increased slightly when the time series representation increased from 1252 clusters to 2920 clusters. However, the simulation time rose rapidly.

The study was mainly focused on a community-scale district heating demand. In the future, the study shall be extended to the municipality-level district heating demand with more deep heat boreholes, while considering the electrical and the DH network. The heat and power losses in the network would somehow affect the benefits obtained from the proposed model.

**Author Contributions:** Conceptualization, A.A.B. and M.L.; methodology, A.A.B.; software, A.A.B., M.P.-K. and A.L.; validation, A.A.B.; formal analysis, A.A.B., M.P.-K. and M.L.; investigation, A.A.B.; writing—original draft preparation, A.A.B.; writing—review and editing, A.L., M.P.-K. and M.L.; supervision, M.L.; project administration, M.L.; funding acquisition, M.L. All authors have read and agreed to the published version of the manuscript.

**Funding:** This research received no external funding.

**Conflicts of Interest:** The authors declare no conflict of interest.

## Nomenclature

| | |
|---|---|
| Sets | |
| $t, T$ | index and set of time steps |
| $n, N$ | index and set of households |
| $i, I$ | index and set of breakpoints for SOS2 variables during heat injection |
| $j, J$ | index and set of breakpoints for SOS2 variables during heat extraction |
| Parameters | |
| $A_1, A_2, A_3$ | Regression coefficients for heat extracted by the DWHP |
| $B_1, B_2, B_3$ | Regression coefficients for the heat extracted by the deep bore hole alone |
| $C_1, C_2, C_3$ | Regression coefficients for the heat injected into the deep bore hole alone |
| $C_n^a$ | Thermal capacitance of indoor air (Wh/°C/m$^2$) |
| $C_n^m$ | Thermal capacitance of building fabric (Wh/°C/m$^2$) |
| $D_i, E_j$ | Breakpoints for SOS2 variables |
| $H_n^m$ | Heat conductance between indoor air and building fabric (W/°C/m$^2$) |
| $H_n^e$ | Heat conductance between external air and indoor air (W/°C/m$^2$) |
| $H_n^y$ | Heat conductance between external air and building fabric (W/°C/m$^2$) |
| $H_n^x$ | Heat conductance between heating air and indoor air (W/°C/m$^2$) |
| $H_n^g$ | Heat conductance between indoor air and ground (W/°C/m$^2$) |
| M | a big number |
| $P_t^w$ | Wind power generation in time step $t$ (W) |
| $P_{\max}^{TS,ch}$ | Maximum charging power of the thermal storage (W) |
| $P_{\min}^{dwhp,e}$, $P_{\max}^{dwhp,e}$ | Minimum and maximum input power of the DWHP during heat extraction, respectively (W) |
| $P_{\min}^{dwhp,i}$, $P_{\max}^{dwhp,,i}$ | Minimum and maximum input power of the DWHP during heat injection, respectively (W) |
| $Q_n^{\max}$ | Rated power of the heating unit of house $n$ (W) |
| $Q_{\max}^{TS,dch}$ | Maximum discharging heat power of the thermal storage (W) |
| $SOC_{\min}^{TS}$ | Minimum allowed state of charge of the thermal storage (Wh) |
| $T^g$ | Ground node temperature (°C) |

| | |
|---|---|
| $T^x$ | Temperature of heating air (°C) |
| $T_n^{set}$ | Indoor set-point temperature of household $n$ (°C) |
| $\alpha$ | Wind curtailment level |
| $\Delta$ | Width of the temperature dead band (°C) |
| $\beta_n^h$ | Annual space heating demand of household $n$ (Wh) |
| $\omega_t$ | Duration of time step $t$ (hours) |
| $\eta^{ch}, \eta^{dch}$ | Charging and discharging efficiency of the thermal storage |
| $u$ | Thermal storage loss coefficient |
| $\phi_{t,n}^{sol}, \phi_{t,n}^{\text{int}}$ | Solar and internal heat gains in house $n$ in time step $t$, respectively (W) |
| Variables | |
| $P_t^{TS,ch}$ | Charging power of the thermal storage (W) |
| $P_t^{dwhp,e}$ | Input power of the DWHP during heat extraction in time step $t$ (W) |
| $P_t^{dwhp,i}$ | Input power of the DWHP during heat injection in time step $t$ (W) |
| $P_t^{wc}$ | Wind power curtailed in time step $t$ (W) |
| $Q_t^{TS,dch}$ | Discharging power of the thermal storage (W) |
| $Q_{t,n}^h$ | Heating power required in house $n$ in time step $t$ (W) |
| $Q_t^{dwhp,e}$ | Heat extracted by the DWHP in time step $t$ (W) |
| $Q_t^{BH,e}$, $Q_t^{BH,i}$ | Heat extracted and injected to the deep bore hole in time step $t$, respectively (W) |
| $SOC_{\max}^{TS}$, $SOC_t^{TS}$ | Maximum capacity and instantaneous capacity of the thermal storage in time $t$, respectively (Wh) |
| $T_{t,n}^a$ | Indoor ambient temperature of house $n$ in time step $t$ (°C) |
| $T_{t,n}^m$ | Building mass temperature of house $n$ in time step $t$ (°C) |
| $u_t$, $v_t$ | Binary variables for the heat extraction and heat injection, respectively |
| $x_{t,n}^h$ | Binary variable for the heating power of household $n$ in time step $t$ |
| $y_{t,i}, z_{t,j}$ | SOS2 variables for calculating the square of continuous variables |
| $^{loss}_t$ | Storage losses of the thermal storage in time step $t$ (Wh) |

## References


1. Arabzadeh, V.; Pilpola, S.; Lund, P.D. Coupling Variable Renewable Electricity Production to the Heating Sector through Curtailment and Power-to-heat Strategies for Accelerated Emission Reduction. *Future Cities Environ.* **2019**, *5*. [CrossRef]
2. Statistics Finland. 2019. Available online: https://www.stat.fi/til/asen/2019/asen_2019_2020-11-19_tie_001_en.html (accessed on 12 January 2021).
3. Arabzadeh, V.; Mikkola, J.; Jasiūnas, J.; Lund, P.D. Deep decarbonization of urban energy systems through renewable energy and sector-coupling flexibility strategies. *J. Environ. Manag.* **2020**, *260*, 110090. [CrossRef] [PubMed]
4. Helin, K.; Syri, S.; Zakeri, B. Improving district heat sustainability and competitiveness with heat pumps in the future Nordic energy system. *Energy Procedia* **2018**, *149*, 455–464. [CrossRef]
5. Lund, P.D.; Skytte, K.; Bolwig, S.; Bolkesjö, T.F.; Bergaentzlé, C.; Gunkel, P.A.; Kirkerud, J.G.; Klitkou, A.; Koduvere, H.; Gravelsins, A.; et al. Pathway Analysis of a Zero-Emission Transition in the Nordic-Baltic Region. *Energies* **2019**, *12*, 3337. [CrossRef]
6. Mikkola, J.; Lund, P.D. Modeling flexibility and optimal use of existing power plants with large-scale variable renewable power schemes. *Energy* **2016**, *112*, 364–375. [CrossRef]
7. Salpakari, J.; Mikkola, J.; Lund, P.D. Improved flexibility with large-scale variable renewable power in cities through optimal demand side management and power-to-heat conversion. *Energy Convers. Manag.* **2016**, *126*, 649–661. [CrossRef]
8. Pilpola, S.; Arabzadeh, V.; Mikkola, J.; Lund, P. Analyzing National and Local Pathways to Carbon-Neutrality from Technology, Emissions, and Resilience Perspectives—Case of Finland. *Energies* **2019**, *12*, 949. [CrossRef]
9. Ali, M.; Ekström, J.; Lehtonen, M. Sizing Hydrogen Energy Storage in Consideration of Demand Response in Highly Renewable Generation Power Systems. *Energies* **2018**, *11*, 1113. [CrossRef]
10. Bashir, A.A.; Pourakbari-Kasmaei, M.; Contreras, J.; Lehtonen, M. A novel energy scheduling framework for reliable and economic operation of islanded and grid-connected microgrids. *Electr. Power Syst. Res.* **2019**, *171*, 85–96. [CrossRef]
11. Bashir, A.; Pourakbari Kasmaei, M.; Safdarian, A.; Lehtonen, M. Matching of Local Load with On-Site PV Production in a Grid-Connected Residential Building. *Energies* **2018**, *11*, 2409. [CrossRef]
12. Bashir, A.A.; Lund, A.; Pourakbari-Kasmaei, M.; Lehtonen, M. Minimizing Wind Power Curtailment and Carbon Emissions by Power to Heat Sector Coupling—A Stackelberg Game Approach. *IEEE Access* **2020**, *8*, 211892–211911. [CrossRef]
13. Helistö, N.; Kiviluoma, J.; Holttinen, H.; Lara, J.D.; Hodge, B.-M. Including operational aspects in the planning of power systems with large amounts of variable generation: A review of modeling approaches. *Wiley Interdiscip. Rev. Energy Environ.* **2019**, *8*, e341. [CrossRef]
14. Frew, B.A.; Jacobson, M.Z. Temporal and spatial tradeoffs in power system modeling with assumptions about storage: An application of the POWER model. *Energy* **2016**, *117*, 198–213. [CrossRef]

15. Asensio, M.; Meneses de Quevedo, P.; Munoz-Delgado, G.; Contreras, J. Joint Distribution Network and Renewable Energy Expansion Planning Considering Demand Response and Energy Storage—Part I: Stochastic Programming Model. *IEEE Trans. Smart Grid* **2018**, *9*, 655–666. [CrossRef]
16. Dvorkin, Y.; Fernandez-Blanco, R.; Kirschen, D.S.; Pandzic, H.; Watson, J.-P.; Silva-Monroy, C.A. Ensuring Profitability of Energy Storage. *IEEE Trans. Power Syst.* **2017**, *32*, 611–623. [CrossRef]
17. De Sisternes, F.J.; Jenkins, J.D.; Botterud, A. The value of energy storage in decarbonizing the electricity sector. *Appl. Energy* **2016**, *175*, 368–379. [CrossRef]
18. Tejada-Arango, D.A.; Domeshek, M.; Wogrin, S.; Centeno, E. Enhanced Representative Days and System States Modeling for Energy Storage Investment Analysis. *IEEE Trans. Power Syst.* **2018**, *33*, 6534–6544. [CrossRef]
19. Brijs, T.; van Stiphout, A.; Siddiqui, S.; Belmans, R. Evaluating the role of electricity storage by considering short-term operation in long-term planning. *Sustain. Energy Grids Netw.* **2017**, *10*, 104–117. [CrossRef]
20. Pineda, S.; Morales, J.M. Chronological Time-Period Clustering for Optimal Capacity Expansion Planning With Storage. *IEEE Trans. Power Syst.* **2018**, *33*, 7162–7170. [CrossRef]
21. Raventós, O.; Bartels, J. Evaluation of Temporal Complexity Reduction Techniques Applied to Storage Expansion Planning in Power System Models. *Energies* **2020**, *13*, 988. [CrossRef]
22. European Commission: Climate and Energy Framework. Available online: https://ec.europa.eu/clima/policies/strategies/2030_en (accessed on 1 December 2020).
23. Bashir, A.A.; Lehtonen, M. Day-Ahead Rolling Window Optimization of Islanded Microgrid with Uncertainty. In Proceedings of the 2018 IEEE PES Innovative Smart Grid Technologies Conference Europe (ISGT-Europe), Sarajevo, Bosnia and Herzegovina, 21–25 October 2018; pp. 1–6.
24. Alimohammadisagvand, B.; Alam, S.; Ali, M.; Degefa, M.; Jokisalo, J.; Sirén, K. Influence of energy demand response actions on thermal comfort and energy cost in electrically heated residential houses. *Indoor Built Environ.* **2017**, *26*, 298–316. [CrossRef]
25. *Decree (1009/2017) of the Ministry of the Environment on Indoor Climate and Ventilation in the New Building, 2017*; Ministry of Housing, Energy and the Environment: Helsinki, Finland, 2017.
26. *Helsingin Geoenergiapotentiaali*; Kaupunkiympäristön julkaisuja 2019:25. Helsinki, 2019. Available online: https://www.hel.fi/static/liitteet/kaupunkiymparisto/julkaisut/julkaisut/julkaisu-25-19.pdf (accessed on 10 October 2020).
27. Wang, Z.; Wang, F.; Liu, J.; Ma, Z.; Han, E.; Song, M. Field test and numerical investigation on the heat transfer characteristics and optimal design of the heat exchangers of a deep borehole ground source heat pump system. *Energy Convers. Manag.* **2017**, *153*, 603–615. [CrossRef]
28. Holmberg, H.; Acuña, J.; Næss, E.; Sønju, O.K. Thermal evaluation of coaxial deep borehole heat exchangers. *Renew. Energy* **2016**, *97*, 65–76. [CrossRef]
29. Lund, A.; Karvinen, T.; Lehtonen, M. Analysis of deep-heat energy wells for heat pump systems. In Proceedings of the 2020 IEEE PES Innovative Smart Grid Technologies Europe (ISGT-Europe), Hague, The Netherlands, 26–28 October 2020; pp. 574–578.
30. Lund, A. *Analysis of Deep-Heat Energy Wells for Heat Pump Systems*; Aalto University: Espoo, Finland, 2019.
31. Ward, J.H. Hierarchical Grouping to Optimize an Objective Function. *J. Am. Stat. Assoc.* **1963**, *58*, 236–244. [CrossRef]
32. Pourakbari Kasmaei, M.; Asensio, M.; Lehtonen, M.; Contreras, J. Trilateral Planning Model for Integrated Community Energy Systems and PV-based Prosumers-A Bilevel Stochastic Programming Approach. *IEEE Trans. Power Syst.* **2019**, *35*, 346–361. [CrossRef]
33. Alahäivälä, A.; Corbishley, J.; Ekström, J.; Jokisalo, J.; Lehtonen, M. A control framework for the utilization of heating load flexibility in a day-ahead market. *Electr. Power Syst. Res.* **2017**, *145*, 44–54. [CrossRef]
34. Ekström, J.; Koivisto, M.; Mellin, I.; Millar, J.; Saarijärvi, E.; Haarla, L. Assessment of large scale wind power generation with new generation locations without measurement data. *Renew. Energy* **2015**, *83*, 362–374. [CrossRef]
35. Finnish Meteorological Institute, Finland. Available online: http://en.ilmatieteenlaitos.fi/ (accessed on 10 October 2020).

MDPI
St. Alban-Anlage 66
4052 Basel
Switzerland
Tel. +41 61 683 77 34
Fax +41 61 302 89 18
www.mdpi.com

*Energies* Editorial Office
E-mail: energies@mdpi.com
www.mdpi.com/journal/energies

www.ingramcontent.com/pod-product-compliance
Lightning Source LLC
Chambersburg PA
CBHW062058220926
51850CB00010B/2816